Super Greens

Gokhan Hacisalihoglu

Super Greens

A Guide to Plants in the Human Context

Springer

Gokhan Hacisalihoglu
Biological Sciences Department
Florida Agricultural and Mechanical
University
Tallahassee, FL, USA

ISBN 978-3-032-17046-0 ISBN 978-3-032-17047-7 (eBook)
https://doi.org/10.1007/978-3-032-17047-7

This Springer imprint is published by the registered company Springer Nature Switzerland AG
The registered company address is: Gewerbestrasse 11, 6330 Cham, Switzerland

Preface

Super Greens 1, 2:

1. Packed with chlorophyll, proteins, minerals, fiber, vitamins, antioxidants, and polyphenols.
2. Amazing in supporting the planet inside and out.

Welcome to Super Greens! Full of modern, relevant examples, *Super Greens* is a new, relevant, exciting, concise, yet comprehensive introduction from agriculture to human health and future sustainability concepts for the reader with no experience in the field. It has user-friendly, effective learning chapters by breaking down complex concepts into practical terms in this growing subject area.

Plants are less often adequately discussed, even though they are a cornerstone of biology. This text fills these gaps with an emphasis on the origins of agriculture, plant anatomy, morphology, reproduction, crop physiology, genetics, horticulture, food, beverage, fiber, and medicinal plants, diets, nutrition, superfoods, and future plants. It also highlights case studies in plants and success tips features for plant biology mastery.

Super Greens' each chapter features a list of student learning outcomes (SLOs), a big picture infographic summary, numerous color figures, review questions, and a set of plant profiles that inspire deeper learning about plants in the human context.

Learners from several levels, including college students, growers, ag advisors, and the general public alike, will greatly enhance their plant biology education with the help of *Super Greens*. With the *Super Greens*' flexible approach, it will help to strengthen their plant knowledge and build new skills to excel in college courses. Incorporating the very latest information and research concerning food crop plants, *Super Greens* balances theoretical and practical knowledge to provide introductory courses with in-depth coverage of plant sciences.

Super Greens is the perfect introduction to the plant science subject, with the content presented in organized learning blocks for mastering central concepts, need-to-know examples for students and professionals who are learning on several levels to advance their education in plant biological sciences, and master the core principles of plants in the human context.

Finally, no prerequisite background is assumed to obtain full benefits from this text. It is my sincerest hope that *Super Greens* serves as a "go to" guide to help us faculty, students, and growers to envision the future plants as plant biology is becoming increasingly important worldwide. I invite the readers to peruse the "Contents" to explore *Super Greens*. Stay in the know with the Super Greens!

Tallahassee, FL, USA G. Hacisalihoglu

Acknowledgments This book is inspired by and dedicated to my parents as well as my daughter Ela to teach me learning every day, unwavering support, with thanks. No one writes a book alone. Firstly, a massive thank you to my publisher Springer Nature for giving me the opportunity to write this book, which was my privilege to complete. I am very grateful to Springer Nature senior editor Dr. Kenneth Teng and the Springer Nature team for their hard work and for providing recommendations and encouragement. I am grateful to the outstanding project coordinator Gifty Priscilla, for providing all-around support and guidance through the writing process.

As ever, I am indebted to all mentors and researchers who continue to study STEM sciences, as well as Keystone Writing Program, Drs. Dik, Cousins, Kochian, Latinwo, Pierre, Thornton, Guner, Sen, Robinson, Soylemezoglu, YTK, RT, Gulec, Momol, Mateeva, Yariz, Olgar, Pendleton, Caglar, Fujiwara, Hyndman, Duke, Armstrong, Blair, Salters, Settles, Han, Lin, Ozturk, Cakmak, Niklas, and my brothers M.E. and A.

This book has been made stronger by all students throughout my 23 years of teaching biology, and of course, to my readers/followers for your generous comments and encouragement.

Finally, lasting thanks to all for growing a greener planet, and enjoy this wonderful journey with Super Greens!

Contents

About the Author

Gokhan Hacisalihoglu (PhD, Cornell) is a Professor of Biology and World-Class Faculty at Florida A&M University, USA, and a Courtesy Professor at FSU. He carries out research into food crops with an emphasis on genetics, bioinformatics, and phenotyping proteins, ionomes, and oils. He led multiple NSF/USDA-funded projects with proven success. He has published widely (60+ articles and author of three books "From Growing to Biology," "Level Up C. Success," and "PBio Concepts LM"). He has taught courses at all levels, including intro and advanced plant biology and majors biology. With over 9600 students taught, he brings unparalleled expertise into delivering high-quality results to help undergraduates to pursue their goals in STEM fields. He is the winner of numerous awards, including national Excellence in HigherEd Award, FAMU Distinguished Researcher, Adv. Teacher of the Year, and Fulbright to Japan. He has held visiting professor positions at Cornell, FSU, Sabanci University, Michigan State University, University of Tokyo, University of Florida, and USDA. He enjoys traveling and photography (and several of his pictures appear in this book).

Find out more at:

Linkedin: www.linkedin.com/in/hacis/

YouTube: https://www.youtube.com/c/DrhaciExplains

Chapter 1
Origins and Agriculture

Origins and Agriculture

Contents

G. Hacisalihoglu, *Super Greens*, https://doi.org/10.1007/978-3-032-17047-7_1

1.1 Big Picture: Quick Read Chapter Infographic

1.2 Student Learning Outcomes (SLOs) of the Chapter

- Understand plant evolution and origins.
- Classify and identify major plant families and species.
- Distinguish between nonvascular and vascular plants.
- Define sexual life cycles of major plant groups.
- Know the top 20 things about plants.

1.3 Concepts of the Chapter

1.3.1 Origins, Plants, and Agriculture

The hierarchy levels from biosphere to organism are as follows (Fig. 1.2):

- UNIVERSE>> > GALAXIES> SOLAR SYSTEM> EARTH>. . . ..
- **BIOSPHERE LEVEL:** Earth's living part [AIR (atmosphere] + LAND [lithosphere] + H2O [hydrosphere]
- **BIOME LEVEL** (several ecosystems): Major life zones (e.g., rainforest)
- **ECOSYSTEM LEVEL** (living + nonliving things): BIOTIC FACTORS (plants, animals) and ABIOTIC (air, soil, H_2O, light, minerals)
- **COMMUNITY LEVEL** (different species group, just biotic)
- **POPULATION LEVEL** (group of same species): Unit of evo.
- **ORGANISM (SPECIES) LEVEL:** Plant

For more on this topic, see Bar-On et al. (2018), Diamond (2002), Massawe et al. (2016), McCouch et al. (2013), Renard and Tilman (2019), Pontzer et al. (2012), and many references therein.

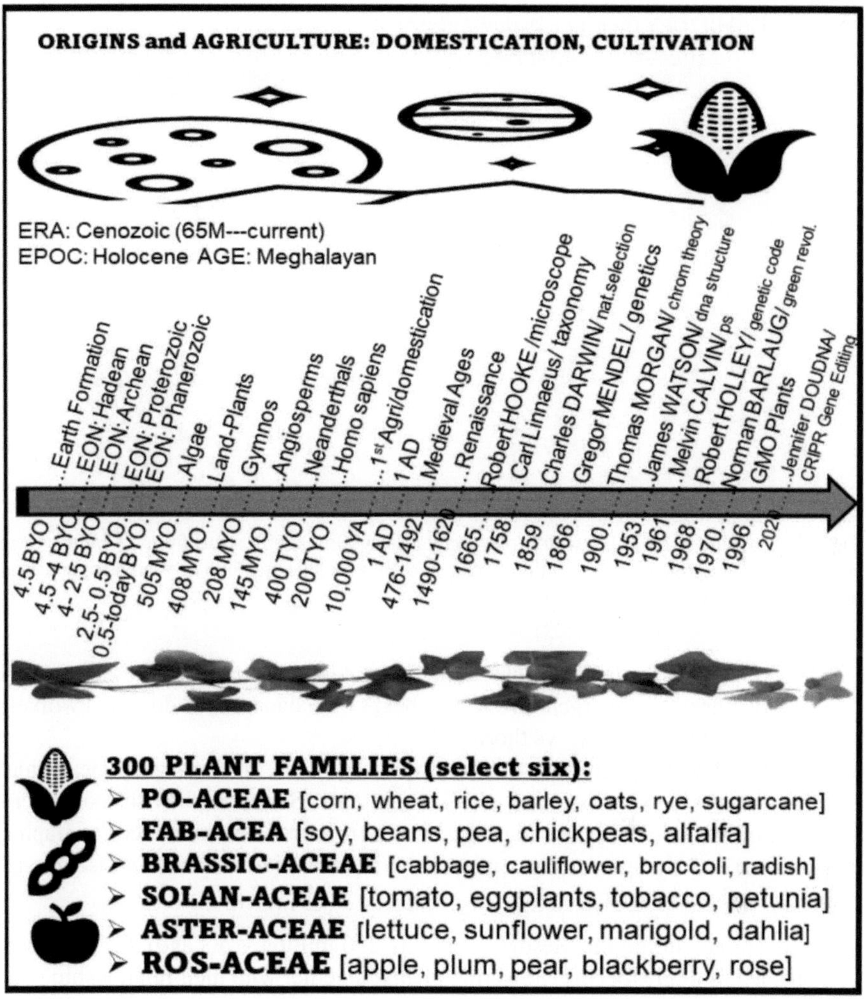

Fig. 1.1 Infographics of this chapter

1.3.2 *Domestication from Wild Plant Ancestors to Modern Crops: 9000+ Years Selection and Cultivation*

Evolution is a change in a population's genetic composition over generations of years. Closest evolutionary relationship also refers to "**most recent common ancestor**." Evolutionary mechanisms include natural selection (survival fitness), mutations (new alleles), gene flow, and genetic drift. How does evolution occur? The two types of evolution are as follows:

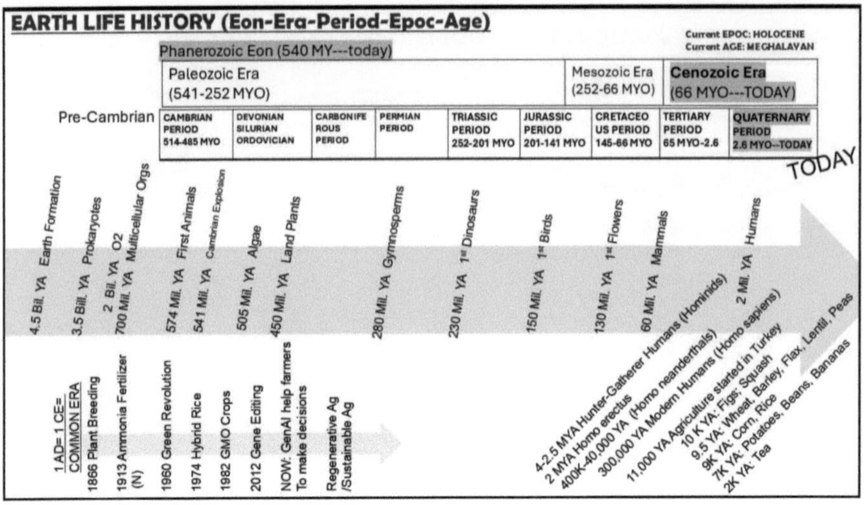

Fig. 1.2 Timeline of Earth life with evolutionary history summary. It spans from 4.5 billion years ago, followed by prokaryotes 3.5 billion years ago, photosynthesis 3 billion years ago, and land plants 450 million years ago

- **MICRO-EVO [mutations, selection]:** Homologous similarity inherited from a common ancestor (divergent evo: Finch beaks of Galapagos; or dogs and wolves; or convergent evo: spines vs thorns)
- **MACRO-EVO [speciation, extinction]:** Large scale through geological time. (e.g., spore plants→ to→seed plants evolution). Some mechanisms of macro evolution include allopatric (e.g., genetic drift, founder effect under geographic isolation) and Sympatric (e.g., polyploidy, reproductive barriers, and isolations/ barriers of reproduction).

For example, during the Mesozoic Era (66–250 million years ago), the evolution of plants led to **angiosperms** (flowering plants) together with their relationship with dinosaurs as well (Fig. 1.3). For more on this topic, see Futuyma and Moreno (1988) and many references therein.

Part 1. Two Billion Years Ago to 10,000 BC: Pre-agricultural Food Plants: This was a hunter-gatherer style. Before agriculture, first humans ate primarily gathered edible wild plants, berries, acorns, wild grains (wild barley, rice, oats), hazelnuts, pine nuts, wild potatoes, dates, and wild grapes. Today, only known last remaining hunter-gatherer tribes are the Hadza people of northern Tanzania (pop: 1300) in Africa, the Inuit of the Arctic, and the Awa-Guaja tribe of the Amazon rainforests (Fig. 1.4).

Part 2. 10,000 BC (12,000 years ago): Neolithic (Agricultural) Revolution—Domestication—Cultivation of Plants for Food in the Fertile Crescent: The domestication of crops has been done by humans over many generations via selection to increase size and decrease toxicity traits among others. As an example, wild small potato ancestors were domesticated 7000 years ago to make them more

Fig. 1.3 The Mesozoic Era is known for being dominated by gymnosperm (e.g., conifers, cycads) plants prehistorically. Many herbivore dinosaurs and their coevolution likely impacted plant evolution to angiosperms, plant biodiversity, as well as plant defense mechanisms. (Image: G. Hacisalihoglu, PhD)

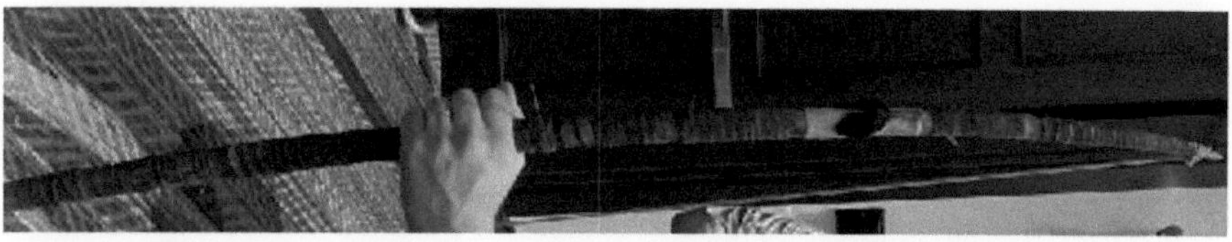

Fig. 1.4 A representative traditional wood bow used by last remaining hunter-gatherers. Hunter-gatherers have no crops and almost no agriculture. (Image: G. Hacisalihoglu, PhD)

useful and edible for humans. In return, this led to a transition to settled agricultural societies from hunter-gatherers. Furthermore, **wheat, barley, lentils, and peas** were domesticated 10,000 years ago, and rice and corn were domesticated 9000 years ago, followed by beans, olives, grapes, dates, and pomegranates.

Furthermore, domestication changed food crop plants. It is estimated that 10% of edible plants are domesticated (2500 out of **25,000 edible plants**). Domesticated plant species changed from wild ancestor plants in many traits, including loss of thorns, loss of bitterness, loss of seeds, smaller seeds, bigger fruits, new color fruits, and climate-adapted varieties, among others. Furthermore, cabbage, broccoli, cauliflower, and kale were all developed from the wild mustard plant single ancestor (*Brassica oleracea*) via selective breeding.

Part 3. 2025 Nation's Food Supply Contributions: In terms of food supply and agricultural production, the leading states are as follows:

- **California:** Provides 11% (=$60 bill.) of nation's food supply (key crops: grapes, almonds, pistachios, olives, figs, walnuts, lettuce, and more)

- **Iowa:** Provides 8% (=$45 bill.) of nation's food supply (key crops: corn and soybeans)
- **Nebraska:** Provides 7% (=$32 bill.) of nation's food supply (key crops: corn, soybeans, and bioethanol)
- **Texas:** Provides 7% (=$30 bill.) of nation's food supply (key crops: cotton, sorghum, and vegetables)
- **Illinois, Minnesota, Indiana:** Corn
- **Kansas (America's breadbasket):** Provides 6% (=$24 bill.) of nation's food supply (key crops: wheat, sorghum, and corn)
- **Florida:** Provides 2% (=$10 bill.) of nation's food supply
- FL is first in sweet corn, cucumbers, oranges, vegetables, floriculture, sugarcane, watermelons, and snap beans
- FL is second in strawberries, peppers, tomatoes (fresh), grapefruits, cabbage, and squash

For more on this topic, see Alexandratos and Bruinsma (2012), Amos et al. (1997), Egli (2017), Farah (2019), GRIN Global (2017), and many references therein.

1.3.3 Biology at Many Dimensions: Cell Sizes, Types, and Development

Have you ever thought about the **microscopic** sizes of **cells, organelles, metabolites**, and their metabolic efficiency? Scales in biology and plant biology include many dimensions (Fig. 1.5). These dimension scales range from micro to meso to macro levels. They also vary in many other dimensions, including cell types (prokaryotic, eukaryotic), **life cycle duration** (annual, biannual, perennial), and development stages (meristem, mature, abscission), among others (Fig. 1.5).

1.3.4 Kingdom: Plantae Classification

Life on Earth and oxygen are made possible due to plants. The four major groups of plant classifications are as follows:

- **NONVASCULARS** (20 K sp): spore-bearing. (e.g., **mosses**).
- **SEEDLESS VASCULARS** (12 K sp): spore-bearing. Plus root. (e.g., **ferns**).
- **GYMNOSPERMS** (1 K sp): naked seeds. (e.g., **pines**).
- **ANGIOSPERMS** (250 K sp): flowering. (e.g., Angiosperms).

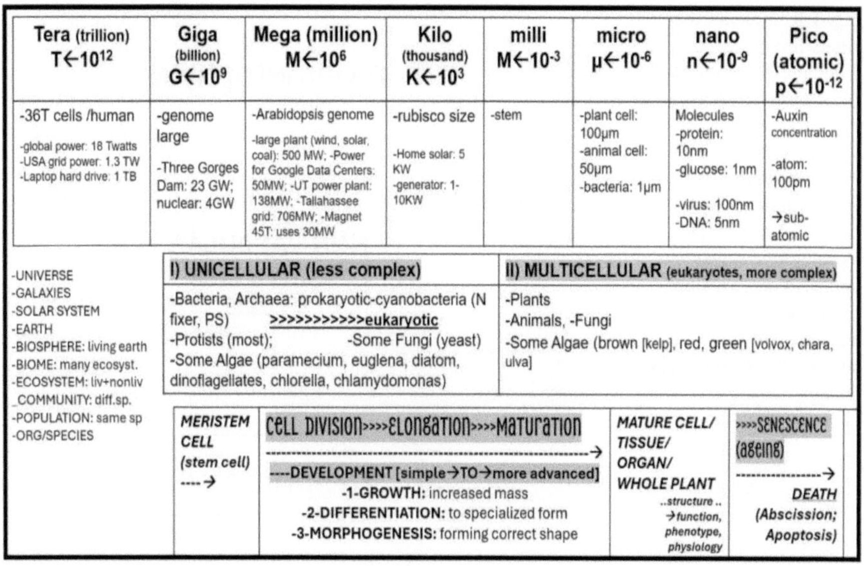

Fig. 1.5 The world of biological dimensions, scales, cell sizes, types, and development stages. The scale ranges from the atomic level to the global ecosystem level

1.3.5 Nonvascular Plants: First Land Plants

First land plants moved to land and benefited from more CO_2, sunlight, and nutrients. Nonvascular plants reproduce with spores (Fig. 1.6). The three major phyla of nonvascular plants are as follows:

1. **Bryophyta** (mosses): e.g., *Mnium, Sphagnum*
2. **Hepatophyta** (liverworts): e.g., *Marchantia*
3. **Anthocerophyta** (hornworts): e.g., *Anthoceros*

All land plant groups have **alternation of generation**, which is the shift between the haploid and diploid phases of plant sexual life cycles. As plants evolved, diploid sporophyte stage became larger as the haploid gametophyte stage became smaller. As a result, nonvascular plants have a larger gametophyte. Gametophytes produce sperm in antheridia organ and **eggs in archegonia organ**. Then, they fuse to produce zygote, which develops sporophyte. Nonvascular plants stay small and near water sources, because their sperm require water to swim for fertilization.

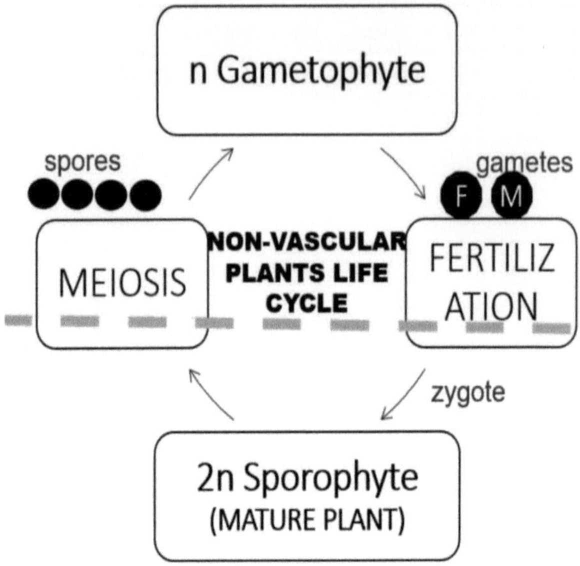

Fig. 1.6 Nonvascular plants life cycle. Their dominant generation is gametophyte. Sporophyte (2n) produces spores via meiosis. Then spores become gametophyte (n), which produce gamete (s) via mitosis. Male reproductive organ (antheridium) produces sperm, while female reproductive organ (archegonium) produces the egg

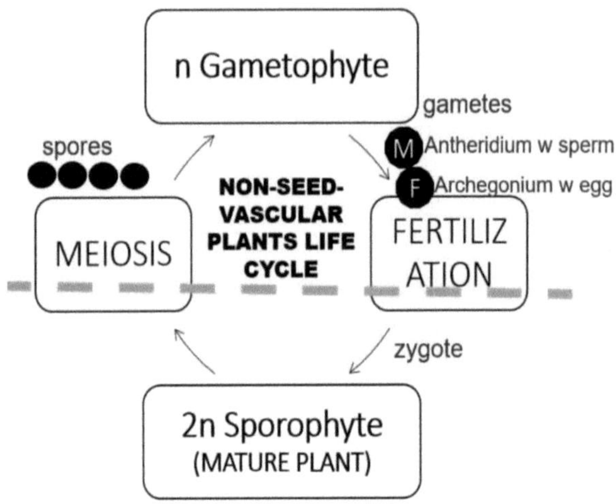

Fig. 1.7 Seedless vascular plants life cycle. Their dominant generation is gametophyte. Male reproductive organ (antheridium) produces sperm, while female reproductive organ (archegonium) produces egg

1.3.6 Seedless Vascular Plants (Vascular = Conducting, Tracheophytes)

Seedless vascular plants reproduce with spores (Fig. 1.7). The four major phyla of seedless vascular plants are as follows:

1. Lycopodiophyta (lycophytes): e.g., *Selaginella*
2. Sphenophyta (horsetails): e.g., *Equisetum*

3. Psilotophyta (whiskferns): e.g., *Psilotum*
4. Pteridophyta (ferns): e.g., *Pteridium, Staghorn, Bracken*

As plants evolved, diploid sporophyte stage became larger as the haploid gametophyte stage became smaller. As a result, vascular plants have a larger sporophyte. Gametophytes produce sperm in antheridia organ and eggs in archegonia organ. Then, they fuse to produce zygote, which develops sporophyte.

1.3.7 Gymnosperms Plants ([Naked]-Seed Plants)

Jurassic period (201 to 65 million years ago) was dominated by gymnosperms and **herbivorous dinosaurs** (**Brachiosaurus**, Apatosaurus, and Stegosaurus). Seed is the reproductive unit of mature fertilized ovule in seed plants (Fig. 1.8). Seed is multicellular, while spore is unicellular (in non-seed plants). The four major phyla of gymnosperm plants are as follows:

1. **Coniferophyta**: e.g., *pine, fir, juniper, yew*
2. **Gnetophyta**: e.g., *Ephedra*
3. **Cycadophyta**: e.g., *Cycas, Zamia, Sago*
4. **Ginkgoophyta**: e.g., *Ginkgo biloba*

As plants evolved, diploid sporophyte stage became larger as the haploid gametophyte stage became smaller. As a result, gymnosperms have a larger dominant sporophyte. Gymnosperms are heterosporous. Pollen grains contain the sperms and are carried via pollination vectors such as wind or animals. **Cones** are reproductive structures.

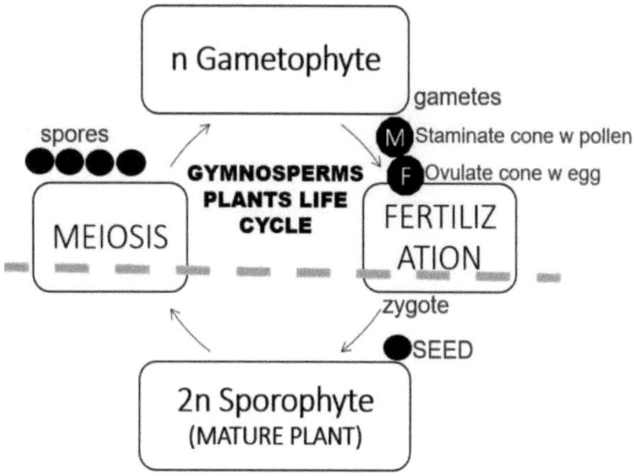

Fig. 1.8 Gymnosperms plant life cycle. Gymnosperms are heterosporous. Their dominant generation is sporophyte. Male reproductive organ (antheridium) produces sperm, while female reproductive organ (archegonium) produces egg

1.3.8 *Angiosperms (Flowering) Plants (Phylum: Anthophyta)*

There are about 250,000 angiosperm species. **Flowers and fruits** are unique to angiosperms. Angiosperm sporophyte produces spores, develops into gametophyte, and produces gametes (Fig. 1.9). Pollen is a spore (microspore) with a potential long-distance transport ability. Angiosperms are heterosporous. Angiosperms' reproductive systems include more advanced flowers and fruits.

The top six most economically important plant families are as follows:

- **POACEAE:** Wheat, rice, corn, barley, and oats (70% of agriculture)
- **FABACEAE:** Beans, soybeans, peas, and much more
- **SOLANACEAE:** Tomato, potato, pepper, and eggplant
- **BRASSICACEAE:** Cabbage, broccoli, mustard, turnips, and *Arabidopsis*
- **CUCURBITACEAE:** Pumpkin, melon, watermelon, and squash
- **APIACEAE:** Carrot, parsnips, celery, parsley, cilantro, and fennel

Anthophyta phylum (two classes: monocots and eudicots). As plants evolved, diploid sporophyte stage became larger as the haploid gametophyte stage became smaller. As a result, angiosperms have a larger dominant sporophyte, which contains smaller gametophyte. Eudicot class plants are the largest flowering plant group.

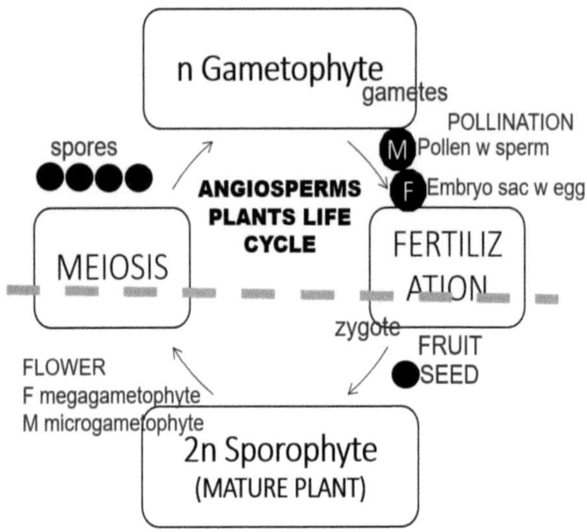

Fig. 1.9 Angiosperms (Flowering) plant life cycle with 250,000 species. Angiosperms are heterosporous. Their dominant generation is sporophyte. In angiosperms, megaspore develops into the megagametophyte, and microspore develops into microgametophyte. There is only a single phylum in angiosperms. Male reproductive organ (stamen) produces sperm, while female reproductive organ (Carpel) produces egg

1.3.9 *Why Study Plants?*

The top 20 things you might not know about plants are as follows:

- There are more than **300,000 + plant species** listed (250 K flowering). Indeterminate growth [continuous] due to meristems. ~2000sp/year newly discovered sp. [e.g., FERNS *Gaga germonatta*].
- Longest Living Tree: **Bristle Cone Pine** in East. Calif. USA (5064 YO; germinated 3050 BC) *Pinus longaeva.*
- **Tallest Plant: Sequoia sempervirens** (Coast Redwood, 330 ft. =100 m); Widest plant: *Sequoia giganteum* (Giant Sequoia, 30 ft. =9 m diameter) (Fig. 1.10).
- **Largest Flower—Single: 1 m wide (Rafflesia arnoldii)**—Compound: 2.5 m tall (*Amorphophallus titanum*). Smallest flower: duckweed (*Wolffia globosa*): 0.3 mm.
- First Vascular Plant (oldest fossil) Known: **Cooksonia** (a Rhyniophyta): 425 MYO.
- Old-Life Span Plants: *Welwitschia* (never dies), *Ginkgo* (1000 YO), Olive (1000 YO).
- Short-Life Span Plants: **Annuals:** Daylily (many 1-day flowers), Morning-Glory (1-day flower), *Arabidopsis thaliana* (40 d).
- Four Things Plants Need: **Water, light, warm temp., nutrients.**
- Medicinal Plants: 70,000 plant species. e.g., willow bark →Aspirin; **French lilac** leaf/flower→Metformin.
- Fastest Growing Plants: **Bamboo** (*Bambusa burmanica*) (90 cm/day).
- Photosynthesis: Plant make their food and oxygen.
- **80,000 Edible Plant spp**.—but 60% from: Wheat, corn, rice.

(A)  (B) 

Fig. 1.10 (**a**) Tallest plants (*Sequoia sempervirens*) (Coast Redwood, 330 ft. =100 m); (**b**) a cross section of redwood tree. (Image: G. Hacisalihoglu, PhD)

- **Center of Origin** (Birthplaces) of Plants: China (rice, soy, cabbage), Türkiye (wheat, rye, peas, olives), Ethiopia (flax, sorghum coffee, okra), Southern MX (maize, beans), Peru (potato, beans), New Guinea (bananas).
- Plant GPS: **Moss grows in tree's north side** in Northern hemisphere.
- Plants grow from top branches; markers stay in same height/place on tree stem.
- Most roots reach 0.6 m; exceptions: Eucalyptus: 40 m.
- Plants are **stationery** (don't move); exceptions: seeds fly away.
- Vascular plants have **xylem** (water-conducting) and **phloem** (food-conducting) tissues.
- World's Hottest **Chili Peppers**: Scoville Heat Units (SHU)→PepperX (2.6 M SHU; Carolina Reaper (2.2 M SHU; Aji Chombo (350 K SHU); Urfa (7500 SHU); Bell pepper (0 SHU).
- **Plant Anatomy** Parts: Root (primary, lateral), stem, leaf, flower (sepal, petal, stamen (anther, microsporangium, pollen), carpel (ovary, ovule, stigma, style, embryo sac, antipodals, central cell, synergids, egg) (Fig. 1.10).

1.3.10 Plant Biodiversity

There are over 300,000 plant species, and each plays a unique role in biological communities. **Plant biodiversity** is important and could save the world one day. There are over 10,000 plant species in danger of potential extinction. Some factors could cause a decline of biodiversity including habitat destruction and invasive species. A list of the most plant biodiverse countries is as follows:

- Brazil: 18% of vascular plants (~35,000 sp.)
- China: 16% of vascular plants (~31,000 sp.)
- Colombia: 10% of vascular plants (~24,000 sp.)
- Peru: 10% of vascular plants (~24,000 sp.)
- Mexico: 8% of vascular plants (~23,000 sp.)
- USA: 6% of vascular plants (~16,000 sp.)
- Türkiye: 3% of vascular plants (~10,000 sp.)

1.3.11 Experimental Model Plants Commonly Used in Research

Model plants are genetically and physiologically similar to other plants and easy to grow. The three major model plants are as follows:

- **DICOT:** *Arabidopsis thaliana* (ARABIDOPSIS)
  Short (36 d) life cycle; well-studied genetically

- **DICOT:** *Pisum sativum* (PEA): N fixation
  Short (56 d) life cycle; high protein
  Also *Medicago truncatula.* Also *Setaria viritis* (C4 model)
- **MONOCOT**: *Zea mays* (CORN)
  Economically very important food crops; great diversity; many mutants
  Also *Oryza sativa* (RICE)

For more on this topic, see Chang et al. (2016) and many references therein.

1.3.12 Other Photosynthetic Species That Live in Water

Although animals do not generally photosynthesize, some may develop symbiotic relationships with algae. The four groups of other photosynthetic species are as follows:

- **CYANOBACTERIA** *(prokaryotic; unicellular; N fixers since 3.5 BY)*→ Phytoplankton; (e.g., *Prochlorococcus marinus*); smallest PS organism (0.7 μm)
- **ALGAE** *(eukaryotic)*→GREEN (e.g., *Chlamydomonas, Ulva);* RED (e.g., *Porypyra, Corallina);* BROWN (e.g., *Fucus; Sargassum);* DIATOMS (e.g., *Fragilaria*); EUGLENOIDS (e.g., *Euglena);* DINOFLAGELLATES (e.g., *Ceratium)*
- **SEA SLUG** *(Elysia chlorotica):* Animal with chloroplast from an algae
- **CORALS:** Animal with symbiosis with dinoflagellates

For more on this topic, see Vital et al. (2023) and many references therein.

1.3.13 Systematics

Systematics is a study of taxonomy (classification) and **phylogeny** (evolutionary relationship) of plants. Phylogenetic trees and cladograms are visual representations of phylogenetic relationships and built on a **common ancestor** between different taxa (Fig. 1.11).

Part 1. Florida Native Plants: These are plant species that have been in Florida for 500 years and well adapted to Florida's climate and soils. Some examples of Florida native plants are:

COREOPSIS (*Coreopsis* spp.): Florida's state wildflower
SAW PALMETTO (*Serenoa repens*)
COONTIE (*Zamia integrifolia*)
DWARF YAUPON HOLLY (*Ilex vomitoria*)

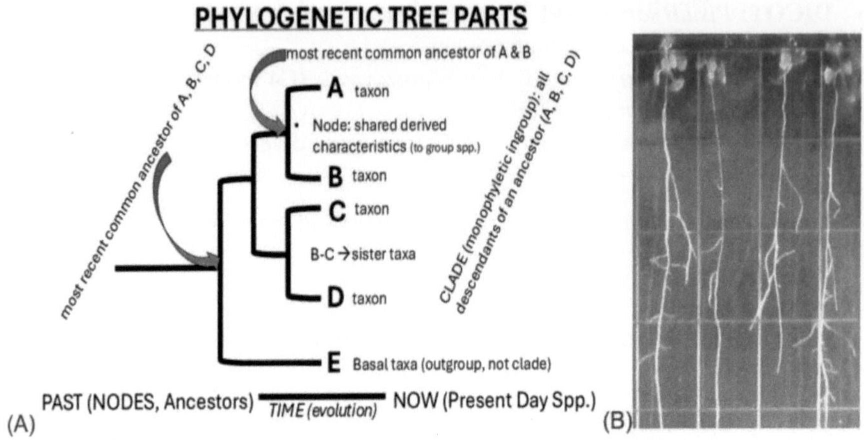

Fig. 1.11 (**a**) Anatomy of a phylogenetic tree; (**b**) *Arabidopsis* (*A. thaliana*) seedlings in agar growing. (Image: G. Hacisalihoglu, PhD)

Plant nomenclature was introduced by Carolus Linnaeus (1758: named and classified ~7700 plant sp.). The major levels of plant biological classification are as follows:

- KINGDOM: PLANTAE (300,000 sp.)
- **PHYLUM/DIVISION:** Anthophyta (250,000 flowering plant)
- CLASS: Dicotyledons (175,000 sp.)
- FAMILY: Brassicaceae (4000 sp.)
- GENUS: *Arabidopsis* (12 sp.)
- SPECIES: *A. thaliana* (thale cress) (Fig. 1.10b)
- VARIETY: Naturally occurring strains [accessions]
- CULTIVAR: Designed by human intervention
- F1-HYBRID: Result of the crossing of two distinct pure sp.
- HEIRLOOMS: Open-pollinated var's w cultural significance

For more on this topic, see Balick and Cox (1996), Eichhorn et al. (2012), Reece et al. (2014), FAO (2017, 2024), FAOSTAT (2023), Hacisalihoglu (2011, 2021, 2022), Harshberger (1896), Hufford et al. (2019), Judd et al. (2015), Levetin and McMahon (2020), Prance et al. (2005), Ridolfi (2019), Schmidt and Cheng (2017), Simpson and Ogorzaly (2013), USDA (2023), USDA-NRCS (2000), Weber et al. (2024), and many references therein.

1.4 Plant Profiles Bingo: How Many Can You Identify?

(P1) 
Abies fraseri

(P2) 
Adenium hybrid

(P3) 
Aechmea fasciata

(P4) 
Agapanthus orientalis

(P5) 
Agave americana

(P6) 
Agave victoriareginae

(P7) 
Alocasia californía

(P8) 
Alocasia macrorrhiza

(P9) 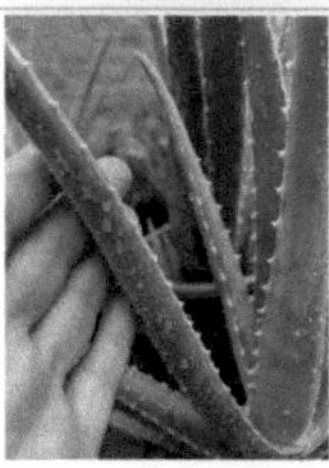
Aloe barbadensis

(P10) 
Aloe ferox

1.5 Takeaway Questions

1. Which major adaptations plants had to transition from aquatic environments to land?
2. How did the domestication of food crops (e.g., wheat, rice) impact early human societies?
3. What is the difference between monocot and dicot plants, and how are they classified in the plant systematics?
4. Why is "cytochrome c" protein commonly used in phylogenetic studies to determine evolutionary relationships of plant species?

1.6 Now You Know the Answers

1. They include the development of vascular tissue (xylem and phloem) for water and inorganic nutrient transport, the evolution of spores and seeds for reproduction on land, a waxy cuticle to prevent water loss, stomata for gas exchange, and roots for anchorage and water absorption.
2. Food crop domestication led to the development of settled communities and from a hunter-gatherer lifestyle to more complex societies, population growth, and civilizations.
3. Monocots and dicots are classified based on their seed structure. Monocots have one cotyledon (seed leaf), parallel leaf veins, and flower parts typically in multiples of three. Dicots have two cotyledons, net-like leaf veins, and flower parts in multiples of four or five. These distinctions are fundamental in plant systematics for identifying and classifying flowering plants (angiosperms).
4. Cytochrome c protein is a highly conserved protein found across many different species of plants, animals, and fungi. Therefore, it is useful for comparing related species with more amino acids they share.

References

Alexandratos N, Bruinsma J (2012) World agriculture towards 2030/2050: the 2012 Revision. Food and Agriculture Organization of the United Nations. ESA Working Paper no. 12–03

Amos A, McCarthy D, Zimmet P (1997) The rising global burden of diabetes and its complications: estimates and projections to the year 2010. Diabetic Med 14:S1–S85

Balick MJ, Cox PA (1996) People, plants, and culture: the science of ethnobotany, New York

Bar-On YM et al (2018) The biomass distribution on earth. Proc Natl Acad Sci U S A 115:6506–6511

Chang C, Bowman JL, Meyerowitz EM (2016) Field guide to plant model systems. Cell 167:325–339

Diamond J (2002) Evolution, consequences and future of plant and animal domestication. Nature 418:700–707. https://doi.org/10.1038/nature01019

Egli DB (2017) Seed biology and yield of grain crops, 2nd edn. CABI International, Boston

Eichhorn SE et al (2012) Raven biology of plants, 7th edn. Macmillan
FAO (2017). The future of food and agriculture: trends and challenges. Food and Agriculture Organization of the United Nations: Rome, Italy
FAO (2024) Food and Agriculture Organization of the United Nations. Current market situation and medium-term outlook 1 August 2025. http://www.fao.org
FAOSTAT (2023) FAO statistical databases. Food and Agriculture Organization of the United Nations: Rome, Italy; https://www.fao.org/faostat
Farah A (2019) Coffee: production, quality and chemistry, 1st edn. The Royal Society of Chemistry, London
Futuyma DJ, Moreno G (1988) The evolution of ecological specialization. Annu Rev Ecol Syst 19: 207–233
GRIN-Global (2017) U.S. National plant germplasm system. GRIN-Global. https://npgsweb.ars-grin.gov/gringlobal/search.aspx?
Hacisalihoglu G (2011) Plant biology concepts and laboratory manual. Kendall Hunt Publishing, Iowa
Hacisalihoglu, G. (2021). From growing to biology: plants 1st edn. Florida State University. 157
Hacisalihoglu G (2022) Level up college success: research, writing, and more key skills for STEM learners. ISBN13: 9798218066260. www.amazon.com/author/drhaci
Harshberger JW (1896) The purposes of ethnobotany. Bot Gaz 21:146–154
Hufford MB et al (2019) Crop biodiversity: an unfinished magnum opus of nature. Annu Rev Plant Biol 70:727–751
Judd WS et al (2015) Plant systematics: a phylogenetic approach, 4th edn. Sinauer Associates, Sunderland
Levetin E, McMahon K (2020) Plants and society, 8th edn. McGraw Hill
Massawe F et al (2016) Crop diversity: an unexploited treasure trove for food security. Trends Plant Sci 21:365–368
McCouch S et al (2013) Feeding the future. Nature 499:23–24
Pontzer H, Raichlen DA, Wood BM, Mabulla AZ, Racette SB, Marlowe FW (2012) Hunter-gatherer energetics and human obesity. PLoS One 7:e40503
Prance GT et al (2005) The cultural history of plants. Routledge, New York
Reece JB et al (2014) Campbell biology, 10th edn. Pearson, Boston
Renard D, Tilman D (2019) National food production stabilized by crop diversity. Nature 571:257–260
Ridolfi R (2019) SDGs for people, planet, and prosperity. http://www.openknowledge.fao.org/
Schmidt BM, Cheng DMK (2017) Ethnobotany: a phytochemical perspective. Wiley
Simpson BB, Ogorzaly MC (2013) Plants in our world: economic botany, 4th edn. McGraw-Hill
USDA (2023) National Agriculture Statistics Service. https://www.nass.usda.gov/
USDA-NRCS (2000) The PLANTS Database. National Resources Conservation Service. https://plants.usda.gov/
Vital XG, Simoes N, Cruz S, Mascaro M (2023) Photosynthetic animals and where to find them: abundance and size of a solar-powered sea slug in different light conditions. Mar Biol 170:154
Weber C, Simnitt S, Wakefield H, Wechsler S, Swearingen B (2024) Fruit and tree nuts outlook: March 2024; United States Department of Agriculture: Washington, DC, USA. https://www.ers.usda.gov/publications/pub-details?pubid=108880

Chapter 2
Plant Micro and Macro Structures

Plant Micro and Macro Structures

Contents

G. Hacisalihoglu, *Super Greens*, https://doi.org/10.1007/978-3-032-17047-7_2

2.1 Big Picture: Quick Read Chapter Infographic

2.2 Student Learning Outcomes (SLOs) of the Chapter

- SLO 02.1: Explain why plant growth is "indeterminate."
- SLO 02.2: Identify and describe 14 key plant cell organelles.
- SLO 02.3: Distinguish between all plastids and pigments.
- SLO 02.4: Identify and describe plant cell types and tissue types.
- SLO 02.5: Identify root system, shoot system, and leaf anatomy (Fig. 2.1).

2.3 Concepts of the Chapter

2.3.1 Plant Cell Level: Grow Your Cells

The **14 organelles** of a rectangular shaped plant cell are nucleus, nucleolus, cell wall, plasma membrane, chloroplasts, mitochondria, endoplasmic reticulum (rough or smooth), Golgi, vacuole, ribosomes, peroxisomes, cytoskeleton, and **plastids**. The organelles specific to plant cells are chloroplast, cell wall, and a central vacuole. Flagella, an animal cell organelle, is found only in plant gametes (Fig. 2.2). Plant cells do not have centrosome (centrioles).

For more on this topic, see Eshel and Beeckman (2013), Evert and Eichhorn (2006), Reeves et al. (2022), USDA-NRCS (2000), Warschefsky et al. (2016), and many references therein.

Part 1: Plastids and Pigments: Pigments are stored within plastids and determine the color of the plastid. Four major plant plastids and their pigments are as follows:

- **Chlorophyll A** [essential primary PS pigment in plants; absorbs light at 430 nm and 663 nm]
- **Chlorophyll B** [accessory PS pigment that assists chlorophyll A; absorbs light at 453 nm and 642 nm]
- **Chromoplast** [pigment→yellow: xanthophyll; orange: carotene; red/purple: anthocyanin; brown: tannins
- **Amyloplast** [non-pigmented]: starch storage

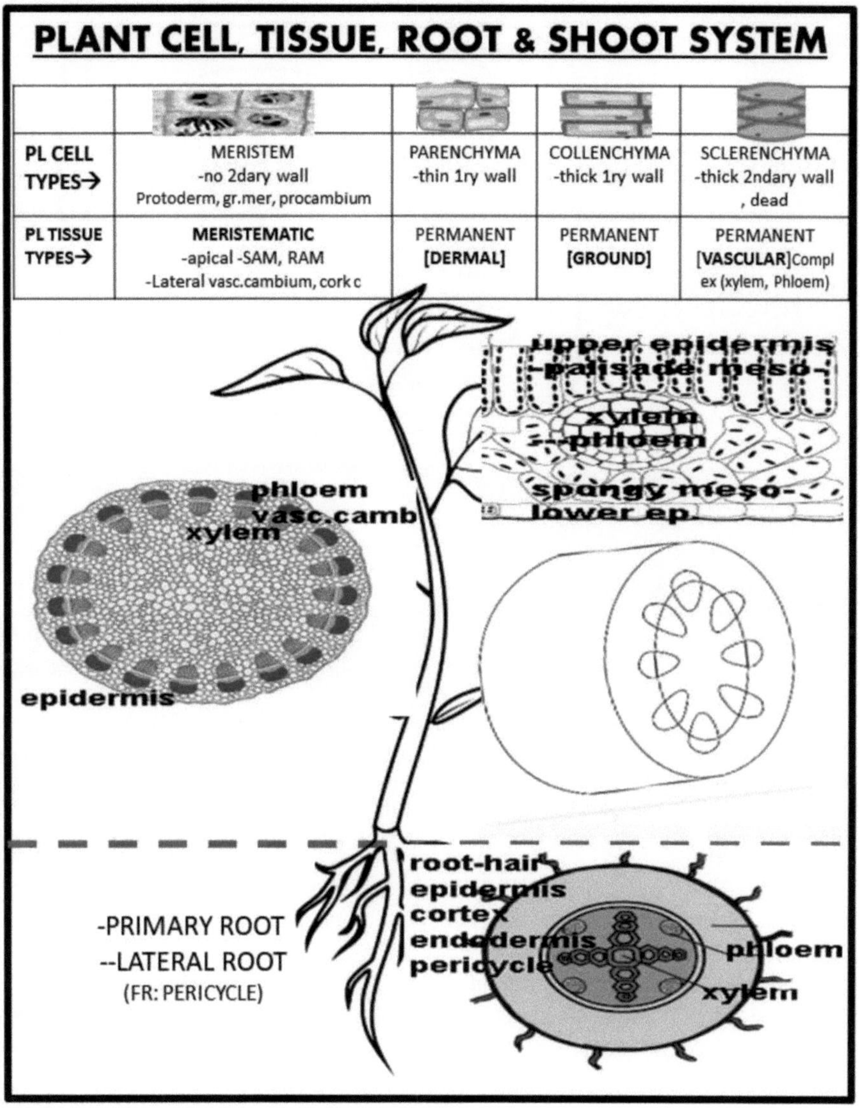

Fig. 2.1 Infographics of this chapter

2.3.2 *Plant Tissue Level Growth*

Meristems with undifferentiated cells are found in shoot tip, root tip, and buds. They become specialized mature plant tissues for plant growth and repair. Therefore, they may be considered similar to animal **stem cells**. Apical meristems (shoot apical

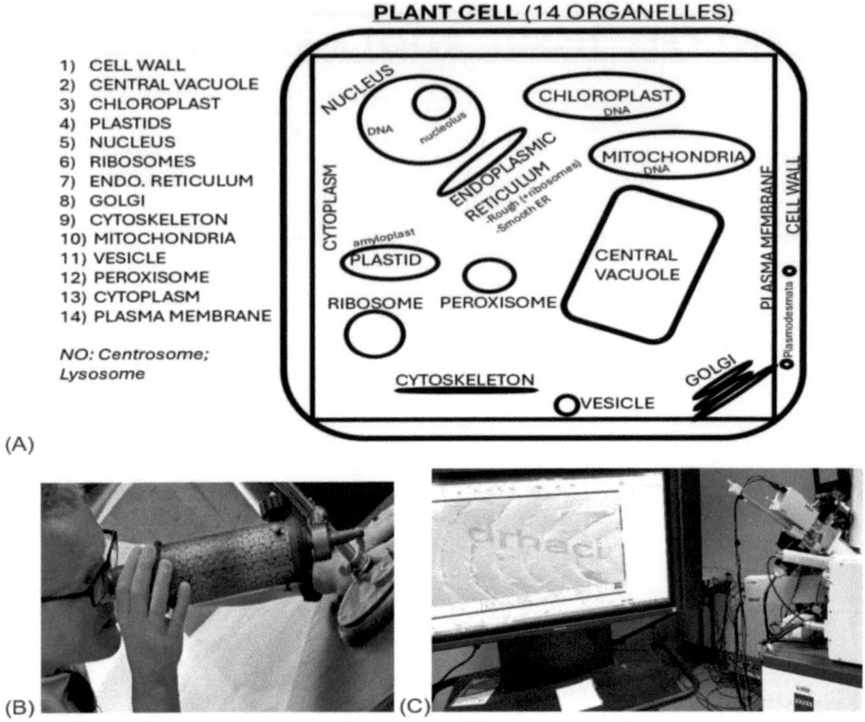

Fig. 2.2 (**a**) All 14 organelles of the plant cell; (**b**) a replica of Robert Hooke's earliest microscope with multiple lenses and oil lamp; (**c**) a scanning electron microscope (SEM) scans an electron beam over to create images and chemical composition analysis of a sample. (Image: G. Hacisalihoglu, PhD)

meristem, SAM, androot apical meristem, RAM) and lateral meristems are two types of meristems in plants (Fig. 2.3).

Plants need light, water, nutrients (inorganic), and temperature (14–30 °C). Measuring growth: length, volume, area, cell number. **Plant development** includes growth and differentiation.

- **Primary Growth: Increase in length** is caused by "apical meristems" (SAM & RAM).
- **Secondary Growth: Widening** is caused by "lateral meristems" (vascular cambium and cork cambium).

Part 1: Plant growth is a combination of the following:

- Cell Division (meristematic)
- Cell Enlargement (expansion)
- Cell Differentiation (develop into specialized)

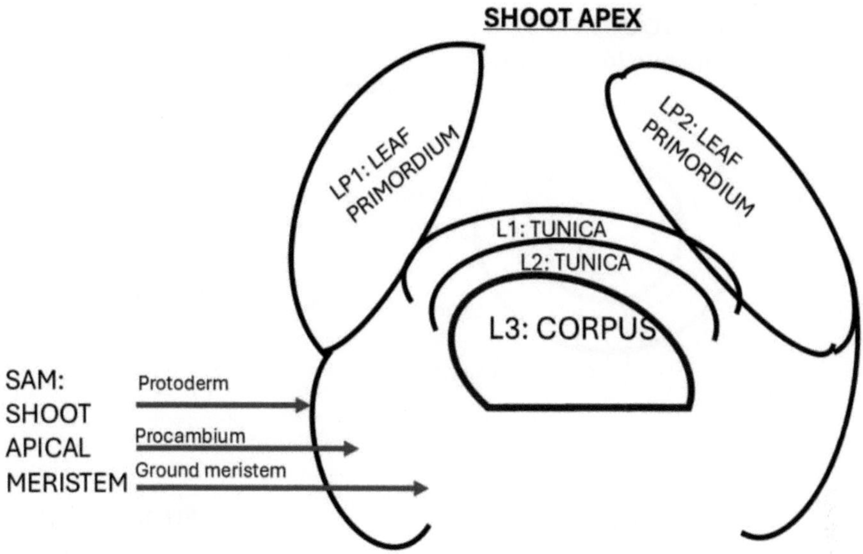

Fig. 2.3 Shoot apex and meristem parts

- Maturation (to form mature plant)
- Morphogenesis (plant shape and structure development process)
- Mature Cell→Senescence→Death

For more on this topic, see Evert and Eichhorn (2006) and many references therein.

2.3.3 Root System

Root is the first plant organ that emerges from the seed embryo. Root anatomy consists of epidermis, cortex, endodermis, pericycle, and the vascular cylinder (stele), which contains primary xylem and primary phloem (Fig. 2.4). Sieve tube cells in phloem lack nucleus at maturity. Casparian strips on endodermis help control the entry of nutrients into xylem. **Taproot** is a long and thick primary root of eudicot class, including *Arabidopsis*, potatoes, and carrots. Water moves in roots via osmosis.

For more on this topic, see Eshel and Beeckman (2013), Evert and Eichhorn (2006), and many references therein.

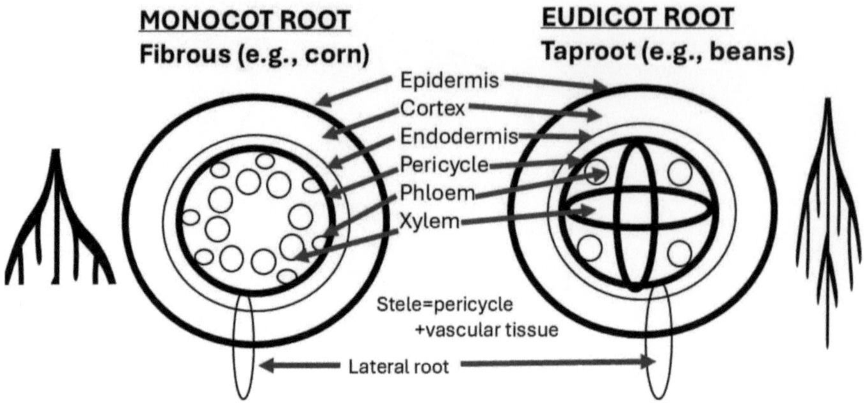

Fig. 2.4 Root anatomy of monocot and eudicot classes

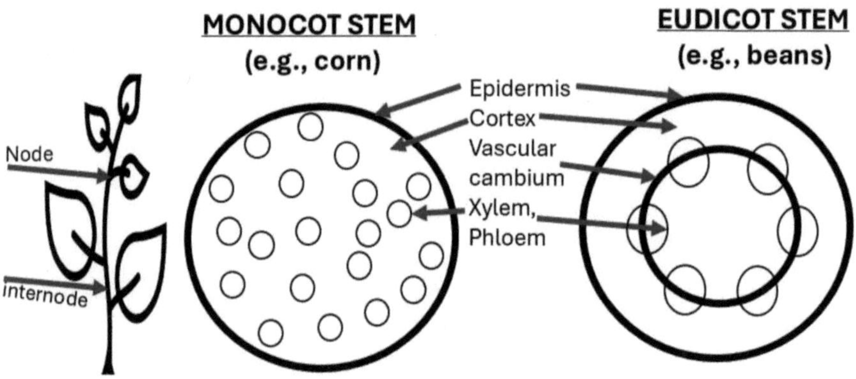

Fig. 2.5 Stem anatomy of monocot and eudicot classes

2.3.4 Shoot System

Plant **shoot system** consists of **stem, leaf, flower, and fruit**. Shoot system grows above ground and supports the plant, transports water and inorganic nutrients from root to leave, and carries out photosynthesis (Fig. 2.5).

Certain plant species develop stem modifications which are homologous to stem. Seven **modified stem** types in plants and representative plant species are as follows:

- Rhizomes (underground) (e.g., ginger)
- Tubers (underground) (e.g., potato) (Fig. 2.6)
- Bulbs (underground) (e.g., onion, lily)
- Corms (underground) (e.g., gladiolus)
- Stolon (ground) (e.g., strawberry)

Fig. 2.6 Potato tubers are underground modified stems and have buds (eyes) for propagation. Tubers are rich starch sources in amyloplasts. (Image: G. Hacisalihoglu, PhD)

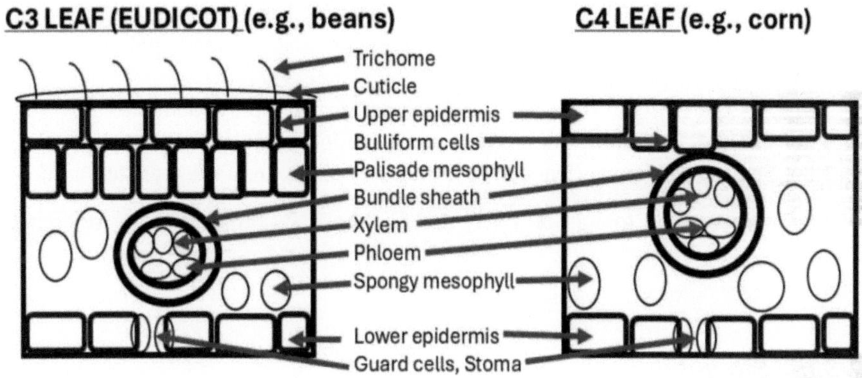

Fig. 2.7 Leaf anatomy of monocot and eudicot classes

- Thorns (aerial) (e.g., rose)
- Cladophylls (aerial) (e.g., cactus)

For more on this topic, see Evert and Eichhorn (2006) and many references therein.

2.3.5 *Leaf (Indeterminate Number of Leaves Depending on Environment)*

Leaf has a specific shape (oval, elliptical, cordate, linear, and more), margin or edge (entire, serrate, lacerate, and more), and venation (parallel, palmate, pinnate, and more).

Leaf anatomy consists of leaf **blade** (leaf surface) petiole (stem attachment), epidermis (outer layer), **mesophyll** with chloroplasts (middle tissue), and vascular bundles (veins for transporting water and nutrients), **stomata** with guard cells (gas exchange) (Fig. 2.7).

For more on this topic, see Evert and Eichhorn (2006) and many references therein.

Nodes bear leaves and leaves often help to identify trees. Leaf patterns used in plant identification are as follows:

Part 1: Leaf Arrangement

- **Simple:** Banana; mango; pear; hibiscus
- **Compound: Pinnately** (hickory; pecan; walnut; honey locust)
- **Compound** (with leaflets)**: Palmately** (buckeye; poison ivy; schefflera)
- **Opposite:** Oak; maple; sycamore (both leaf and branches)
- **Alternate:** Beech; elm; willow (both leaf and branches)
- **Whorled:** Nerium
- **Evergreen:** Conifers (Pine; spruce; fir)
- **Deciduous:** Oak; maple; American beech; ash; elm; birch

Part 2: Leaf Shape: Round, linear; elliptical; heart shaped; palmate; lobed; whorled, lance shaped

Part 3: Leaf Margin: (edge): Entire; lobate; serrate; spiny; undulate

Part 4: Leaf Venation: (veins): Arcuate; parallel; pinnate; palmate

Part 5: Leaf Division: (leaflets): Bifoliate, trifoliate, digitate, pinnate, bipinnate

For more on this topic, see Evert and Eichhorn (2006) and many references therein.

2.3.6 Stomata

Stomata are the tiny pores surrounded by guard cells on leaf surface for water transpiration and gas exchange. The physiological events associated with stomata are as follows:

Part 1: What signal opens stomata?

Light; low CO_2 levels; water in guard cells; K^+ in guard cells.

What happens when stomata open?

CO_2 enters; O_2 exits; water exits (Fig. 2.8).

What signal closes stomata?

Night; high CO_2 levels; water loss; high temperatures; ABA.

For more on this topic, see Willmer and Fricker (1996), Evert and Eichhorn (2006), and many references therein.

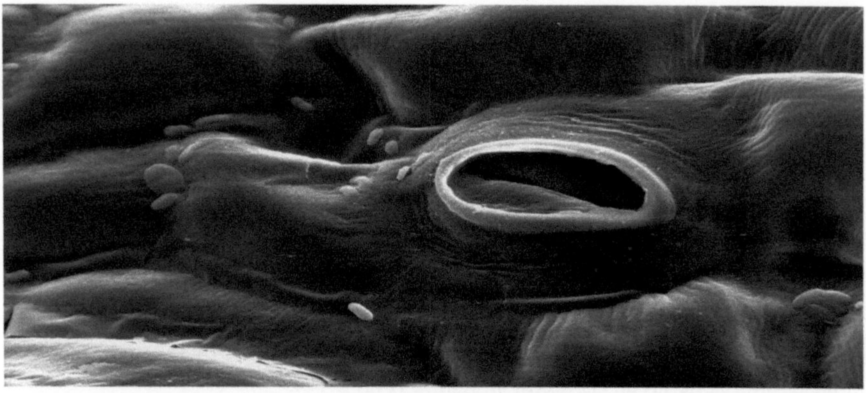

Fig. 2.8 Open stomata are due to water and potassium influx into guard cells in the leaf epidermis. Open stomata also allow gas exchange for photosynthesis pathways

(A) (B)

Fig. 2.9 Representative modified leaves. (**a**) Cactus spines are modified leaf; (**b**) rose thorns are modified stem. They both provide protection against herbivores and water loss reduction

2.3.7 *Modified Leaf Types*

Certain plant species develop leaf modifications which are **homologous** to leaf. Five **modified leaf types** in plants and representative plant species are as follows:

- Spines (e.g., cactus) (Fig. 2.9)
- Needles (e.g., pine)
- Tendrils (e.g., grapes)
- Bracts (e.g., poinsettia)
- Carnivorous traps (e.g., pitcher plant)

2.3.8 *Why Do Leaves Change Colors (Senescence)?*

Certain perennial plant species under certain cooler climate conditions change leaf colors due to **breakdown of chlorophyll** (green) in the **Fall season**. Moreover, other pigments that become visible in the Fall during the senescence are as follows:

- GOLDEN YELLOW: Hickory; aspen; yellow poplar; ash; sycamore
- RED: Dogwood; maples; oaks; sweetgum; persimmons; plum
- ORANGE: Sugar maple; hazelnut
- BROWN: Some oaks

In deciduous plants, **abscission** is the physiological process of shedding their leaf at abscission zone layer via auxin signaling (Fig. 2.10).

For more on this topic, see Balick and Cox (1996), Eichhorn et al. (2012), Reece et al. (2014); FAO (2017, 2024), FAOSTAT (2023), Hacisalihoglu (2011, 2021), Harshberger (1896), Prance et al. (2005), Ridolfi (2019), Schmidt and Cheng (2017), Simpson and Ogorzaly (2013), USDA (2023), USDA-NRCS (2000), Weber et al. (2024), and many references therein.

Fig. 2.10 The degradation of chlorophyll causes assorted colors (yellow, orange, brown, and red) during leaf senescence

2.4 Plant Profiles Bingo: How Many Can You Identify?

(P11) *Aloe striata*

(P12) *Anethum graveolens*

(P13) *Anthirrhinum majus*

(P14) *Anthurium ceranatum*

(P15) *Aspidistra elatior*

(P16) *Calathea lancifolia*

(P17) *Calluna vulgaris*

(P18) *Camelia sinensis*

(P19) *Capsicum annuum*

(P20) *Carica papaya*

2.5 Takeaway Questions

1. Which root zone do root hairs got formed?
2. What are the two photosynthetic cells in C4 plants?
3. Which angiosperm plants have netted venation?
4. Why do leaves change color in the Fall?

2.6 Now You Know the Answers

1. Maturation zone
2. Mesophyll cells and bundle sheath cells
3. Eudicot class plants
4. Breakdown of green chlorophyll

References

Balick MJ, Cox PA (1996) People, plants, and culture: the science of ethnobotany, New York

Eichhorn SE et al (2012) Raven biology of plants, 7th edn. Macmillan

Eshel A, Beeckman T (2013) Plant roots: the hidden half. CRC Press, Boca Raton

Evert RF, Eichhorn SE (2006) Esau's plant anatomy: meristems, cells, and tissues of the plant body: their structure, function, and development. Wiley

FAO (2017) The future of food and agriculture: trends and challenges. Food and Agriculture Organization of the United Nations, Rome

FAO (2024) Food and Agriculture Organization of the United Nations. Current market situation and medium-term outlook 1 August 2025. http://www.fao.org

FAOSTAT (2023) FAO statistical databases. Food and Agriculture Organization of the United Nations: Rome, Italy; https://www.fao.org/faostat

Hacisalihoglu G (2011) Plant biology concepts and laboratory manual. Kendall Hunt Publishing, Iowa

Hacisalihoglu G (2021) From growing to biology: plants, 1st edn. Florida State University, p 157

Harshberger JW (1896) The purposes of ethnobotany. Bot Gaz 21:146–154

Prance GT et al (2005) The cultural history of plants. Routledge, New York

Reece JB et al (2014) Campbell biology, 10th edn. Pearson, Boston

Reeves G et al (2022) Monocotyledonous plants graft at the embryonic root–shoot interface. Nature 602:280–286

Ridolfi R (2019) SDGs for people, planet, and prosperity. http://www.openknowledge.fao.org/

Schmidt BM, Cheng DMK (2017) Ethnobotany: a phytochemical perspective. Wiley

Simpson BB, Ogorzaly MC (2013) Plants in our world: economic botany, 4th edn. McGraw-Hill

USDA (2023) National Agriculture Statistics Service. https://www.nass.usda.gov/

USDA-NRCS (2000) The PLANTS database. National Resources Conservation Service. https://plants.usda.gov/

Warschefsky EJ et al (2016) Rootstocks: diversity, domestication, and impacts on shoot phenotypes. Trends Plant Sci 21:418–437

Weber C, Simnitt S, Wakefield H, Wechsler S, Swearingen B (2024) Fruit and tree nuts outlook: March 2024; United States Department of Agriculture: Washington, DC, USA. https://www.ers.usda.gov/publications/pub-details?pubid=108880

Willmer CM, Fricker M (1996) Stomata (topics in plant functional biology), 2nd edn. Chapman & Hall, London

Chapter 3
Plant Reproduction

Plant Reproduction

Contents

G. Hacisalihoglu, *Super Greens*, https://doi.org/10.1007/978-3-032-17047-7_3

3.1 Big Picture: Quick Read Chapter Infographic

3.2 Student Learning Outcomes (SLOs) of the Chapter

- Understand plant reproduction systems.
- Identify flower structures.
- Distinguish between fruit types.
- Identify and understand flower, fruit, and seed and their anatomy.
- Understand plant secondary growth (Fig. 3.1).

3.3 Concepts of the Chapter

3.3.1 Flowers

What is a flower? **Flowers** consist of four major parts as follows: Sepals (green), **petals (colorful for pollinator vectors)**, stamens (male with pollens), and **carpel** (female with ovary and ovules) (Figs. 3.2 and 3.3). While ovary epigynous refers to inferior or ovary below (e.g., apple or pumpkin), ovary hypogynous refers to superior or ovary above (e.g., beans).

For more on this topic, see Egli (2017), Eichhorn et al. (2012), Reece et al. (2014), and many references therein.

3.3.2 Pollen Types

Pollens (aka microspores or microgametophyte) are carried by pollination vectors (bees, birds, other animals, or wind). Pollination refers to transferring of pollen by a pollination vector from anther (male) to stigma (female). While bee flowers are yellow, white, or blue color and sweet, bird flowers are red, orange, and large. Moreover, wind flowers consist of plant species such as cereals (e.g., corn) and walnuts.

Pollen counts, the number of pollens per m^3 air, are reported to assist individuals with **allergies** to manage their potential allergy symptoms (Fig. 3.4).

For more on this topic, see Levetin and Van de Water (2003) and many references therein.

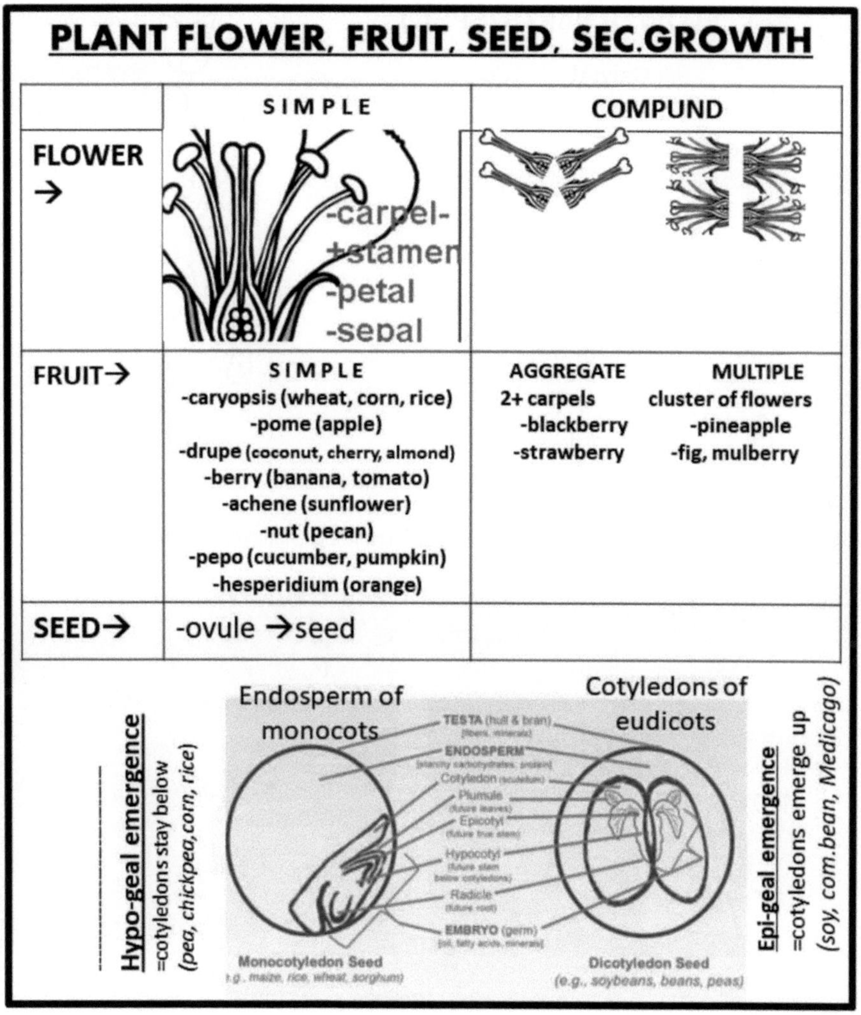

Fig. 3.1 Infographics of this chapter

3.3.3 Mechanisms for Preventing Self-Fertilization

The plant mechanisms to prevent self-fertilization (inbreeding) to increase fitness are as follows:

- **Self-Incompatibility:** Rejection of self-pollen by stigma (no fertilization occurs), e.g., Brassicaceae family
- **Monoecious:** Both male and female separate flowers on the same plant, e.g., corn, oak, squash

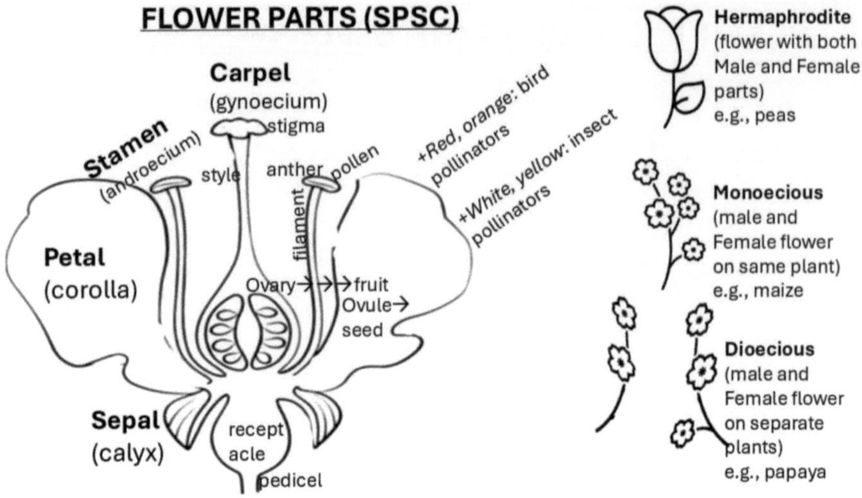

Fig. 3.2 Flower parts and flower types. S (sepal), P (petal), S (stamen), C (carpel)

Fig. 3.3 (**a**) Cherry blossom has long petioles and flower petals with pointy tips plus a notch; (**b**) plum blossom has very short petiole and flower petals with round tips. (Image: G. Hacisalihoglu, PhD)

- **Dioecious:** Male and female flowers born on separate plants, e.g., papaya, asparagus, kiwi, ginkgo, spinach, hemp, poplar, willow

3.3.4 Fruits and True Fruit Types

Fruits (aka mature ovaries) develop after fertilization and contain and protect seeds. The 14 major true fruit type classifications are as follows:

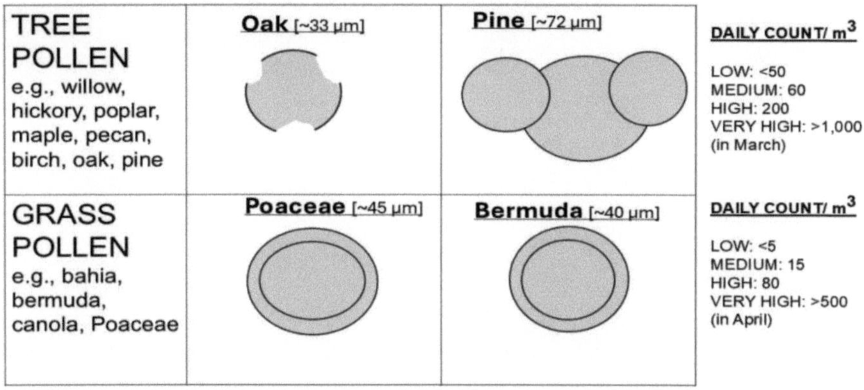

Fig. 3.4 Pollen types and counts for tree pollens and grass pollens

Accessory (fruit contains receptacle)**:** Strawberry, apple, pear
Achene: Buttercup, buckwheat, quinoa
Aggregate (multiple ovaries)**:** Strawberry, blackberry, magnolia
Berry (true berries)**:** Tomato, grape, blueberries, banana, eggplants, avocado, persimmon
Capsule: Cotton, poppy, orchids, willow
Caryopsis: Poaceae family: corn, rice, wheat, oats, rye
Drupe: Peach, cherry, plum, mango, coconut, pistachio, apricot, olive
Hesperidium (subtype berries)**:** Orange, lemon, kumquat
Legume: Beans, pea, soybeans
Multiple (several fused flowers)**:** Fig, pineapple, strawberry
Nut (true nuts)**:** Oak, hazelnut, walnut, pecans (Fig. 3.5)
Pepo (subtype berries)**:** Squash, watermelon, pumpkin, cucumber
Pome: Apple, pear, quince, loquats
Silique: Brassicaceae: cabbage, mustard

For more on this topic, see Janick and Paull (2008), Hacisalihoglu (2021), and many references therein.

3.3.5 Seeds

Seeds, mature **fertilized ovule** with plant embryo and food reserves, consist of the following parts: embryo (tiny plant), endosperm (nutrient rich), testa (seed coat); the embryo itself contains a radicle (primary root), a plumule (embryonic shoot), and one or more cotyledons (seed leaves), which provide nourishment during germination. While **endosperm (3n)** is rich in starch and a good food source, **embryo (2n)** develops into new plant parts such as radicle (baby root), hypocotyl (baby stem), and epicotyl (baby leaf and other upper plant parts (Fig. 3.6).

Fig. 3.5 Representative fruit types. (**a**) Hazelnuts are botanically nut fruit type; (**b**) pumpkins are botanically pepo fruit type. (Image: G. Hacisalihoglu, PhD)

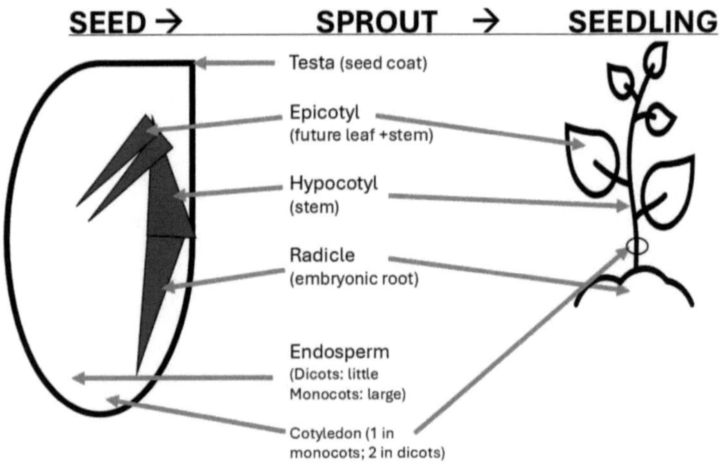

Fig. 3.6 Seed anatomy and seedling morphology parts

For more on this topic, see Hacisalihoglu (2007, 2011, 2021, 2022), Scanlon and Tacas (2009), Tanksley and McCouch (1997), Hacisalihoglu and Ross (2010), Taylor et al. (1998), and many references therein.

3.3.6 Secondary Growth: The Age of Trees

Part 1: WOOD (Secondary Xylem): Secondary growth, a thickening of stem and roots in woody plants, is a growth in girth produced by two lateral meristems in perennial dicots: (i) **vascular cambium** and (ii) **cork cambium**.

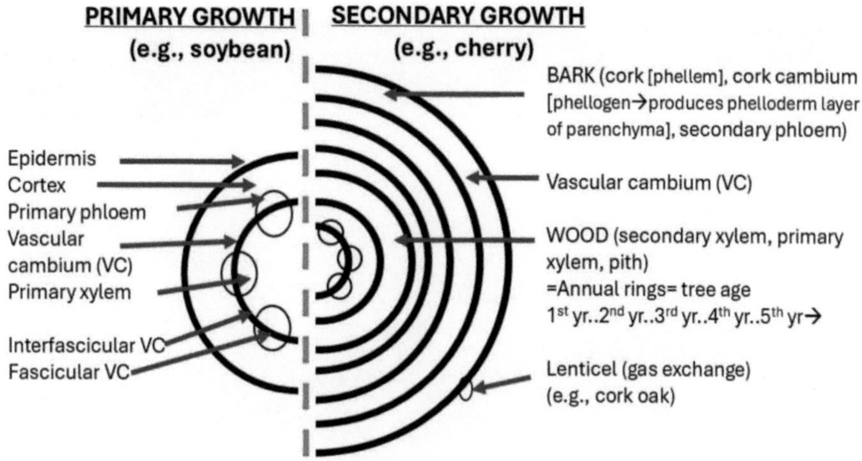

Fig. 3.7 Secondary growth of wood and bark from lateral meristems in plants

Wood is an inner secondary xylem tissue. Bark is the outer layer replaced epidermis and contains lenticels that are small pores with a function of gas exchange (Fig. 3.7).

Vascular cambium (VC): Vascular cambium, a secondary meristem, has fascicular and interfascicular sections for forming secondary tissue. Perennial dicots have secondary growth, while palm trees get thickened through enlargement of parenchyma cells. Vascular cambium is composed of the following two kinds of cells:

- Ray initials of VC: Short, small
- Fusiform initials of VC: Long, narrow

Cork cambium (Phellogen): Cork cambium, also a secondary meristem, forms majority of bark tissue.

Bark = Periderm [phellem (cork) + phellogen (CC) + phelloderm (secondary cortex)] + phloem (secondary). Epidermis in woody plants is replaced by periderm tissue.

Part 2: Dendrochronology: It is a science branch of tree ring dating. **Dendrochronology** also analyzes patterns and density of the **tree rings** to help understand past climate, major natural events, as well as artifacts of ancient wooden structures. As an example, larger rings may suggest favorable environments, while a narrower ring indicate potential drought in that particular year (Fig. 3.8).

For more on this topic, see Rowell (2005), Speer (2010), Stokes and Smiley (1968), and many references therein.

Part 3: Cork (Phellen): Corks are harvested from **cork oak trees** (*Quercus suber*) every 9 years starting at the age 25 until 200 (Fig. 3.9). Cork harvesting process involves separating the thick outermost layer of the tree bark and extracting it from tree. The cork harvesting does not cause tree harm and regrow in 9 years for the following harvest. One cork tree could give enough cork for 100,000 corks in each harvest.

Fig. 3.8 Dendrochronology studies the age of trees, tree rings (secondary xylem), and wood (early wood and late wood). (Image: G. Hacisalihoglu, PhD)

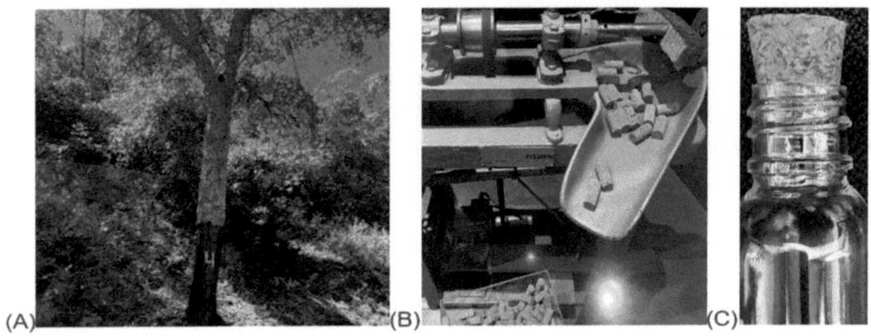

Fig. 3.9 (**a**) Cork oak tree in Portugal. Marking on the tree indicates the year the cork was harvested (4 for the year 2024). The next cork harvest will be in 9 years (2033); (**b**) the puncher cutting the corks; (**c**) Cork in usage. (Image: G. Hacisalihoglu, PhD)

For more on this topic, see Balick and Cox (1996), Eichhorn et al. (2013), Reece et al. (2014), FAO (2017, 2024), FAOSTAT (2023), Hacisalihoglu (2011, 2021, 2022), Harshberger (1896), Judd et al. (2015), Levetin and McMahon (2020), Prance et al. (2005), Ridolfi (2019), Schmidt and Cheng (2017), Simpson and Ogorzaly (2013), USDA (2023), USDA-NRCS (2000), Weber et al. (2024), and many references therein.

3.4 Plant Profiles Bingo: How Many Can You Identify?

(P21)  *Carnegia gigantea*	(P22)  *Carya ilinoinensis*
(P23)  *Celosia plumosa*	(P24)  *Chamelauchium uncinatum*
(P25)  *Chlorophytum comosum*	(P26)  *Cinnamomum verum*
(P27)  *Cocos nucifera*	(P28)  *Coleus scutellarioides*
(P29)  *Copia poakrainziana*	(P30)  *Crassula lycopodioides*

3.5 Takeaway Questions

1. What initiates secondary growth in woody plants?
2. What is an accessory fruit, and can you give an example?
3. What is a multiple fruit, and can you give an example?
4. How do flowers attract pollinators?
5. What adaptations do wind-pollinated flowers have, and can you give examples?

3.6 Now You Know the Answers

1. Secondary growth is initiated by the activity of the vascular cambium and cork cambium. The vascular cambium produces secondary xylem (wood) and secondary phloem, while the cork cambium produces bark.
2. Accessory fruit is a fruit in which additional parts of the flower, apart from the ovary, contribute to the structure of the fruit, e.g., strawberry.
3. Multiple fruits develop from a cluster of flowers (inflorescence), e.g., pineapples.
4. Petal colors attract bees, butterflies, and birds. Scent: Draws insects like moths and bees. Nectar and pollen: Serve as food sources for pollinators.
5. Lack of nectar, petals (or small), large stamens, and stigmas, e.g., corn, rice, pecans, and pistachios.

References

Balick MJ, Cox PA (1996) People, plants, and culture: the science of ethnobotany, New York

Egli DB (2017) Seed biology and yield of grain crops, 2nd edn. CABI International, Boston

Eichhorn SE et al (2012) Raven biology of plants, 7nd edn. Macmillan

Eichhorn SE et al (2013) Raven biology of plants, 7nd edn. Macmillan

FAO (2017) The future of food and agriculture: trends and challenges. Food and Agriculture Organization of the United Nations, Rome, Italy

FAO (2024) Food and Agriculture Organization of the United Nations. Current market situation and medium-term outlook 1 August 2025. http://www.fao.org

FAOSTAT (2023) FAO Statistical Databases; Food and Agriculture Organization of the United Nations: Rome, Italy; https://www.fao.org/faostat

Hacisalihoglu G (2007) Germination characteristics of three warm-season Turfgrasses subjected to Matriconditioning and aging. HortTechnology 17(4):480–485

Hacisalihoglu G (2011) Plant biology concepts and laboratory manual. Kendall Hunt Publishing, Iowa

Hacisalihoglu G (2021) From growing to biology: plants, 1st edn. Florida State University, p 157

Hacisalihoglu G (2022) Level up college success: research, writing, and more key skills for STEM learners. ISBN13: 9798218066260. www.amazon.com/author/drhaci

Hacisalihoglu G, Ross Z (2010) The influence of priming treatment on germination and soil emergence of nonaged and aged annual ryegrass seeds. Seed Sci Technol 38(1):214–217. https://doi.org/10.15258/sst.2010.38.1.21

Harshberger JW (1896) The purposes of ethnobotany. Bot Gaz 21:146–154

Janick J, Paull RE (2008) Encyclopedia of fruit and nuts, 1st edn. CABI, Wallingford, pp 622–624
Judd WS et al (2015) Plant systematics: a phylogenetic approach, 4th edn. Sinauer Associates, Sunderland
Levetin E, McMahon K (2020) Plants and society, 8nd edn. McGraw Hill
Levetin E, Van de Water PK (2003) Pollen count forecasting. Immunol Allergy Clin N Am 23:423–442
Prance GT et al (2005) The cultural history of plants. Routledge, New York
Reece JB et al (2014) Campbell biology, 10th edn. Pearson, Boston
Ridolfi R (2019) SDGs for people, planet, and prosperity. http://www.openknowledge.fao.org/
Rowell RM (2005) Handbook of wood chemistry and wood composites. CRC Press, p 487
Scanlon MJ, Tacas EM (2009) Kernel biology. In: Handbook of maize: its biology, pp 121–143
Schmidt BM, Cheng DMK (2017) Ethnobotany: a phytochemical perspective. Wiley
Simpson BB, Ogorzaly MC (2013) Plants in our world: economic botany, 4th edn. McGraw-Hill
Speer JH (2010) Fundamentals of tree-ring research. University Arizona Press, Tucson
Stokes MA, Smiley TL (1968) An introduction to tree-ring dating. University of Chicago Press, Chicago
Tanksley SD, McCouch SR (1997) Seed banks and molecular maps: unlocking genetic potential from the wild. Science 277:1063–1066
Taylor AG et al (1998) Seed enhancements. Seed Sci Res 8:245–256
USDA (2023) National Agriculture Statistics Service. https://www.nass.usda.gov/
USDA-NRCS (2000) The PLANTS Database. National Resources Conservation Service. https://plants.usda.gov/
Weber C, Simnitt S, Wakefield H, Wechsler S, Swearingen B (2024) Fruit and tree nuts outlook: March 2024; United States Department of Agriculture: Washington, DC, USA. https://www.ers.usda.gov/publications/pub-details?pubid=108880

Chapter 4
Plant Physiology

Plant Physiology

Contents

G. Hacisalihoglu, *Super Greens*, https://doi.org/10.1007/978-3-032-17047-7_4

4.1 Big Picture: Quick Read Chapter Infographic

4.2 Student Learning Outcomes (SLOs) of the Chapter

- Apply the best practices for learning overall plant metabolism.
- Describe biochemical processes of photosynthesis.
- Distinguish between active and passive transport in plants.
- Analyze how external factors affect plant function.
- Identify different plant pigments and their functions (Fig. 4.1).

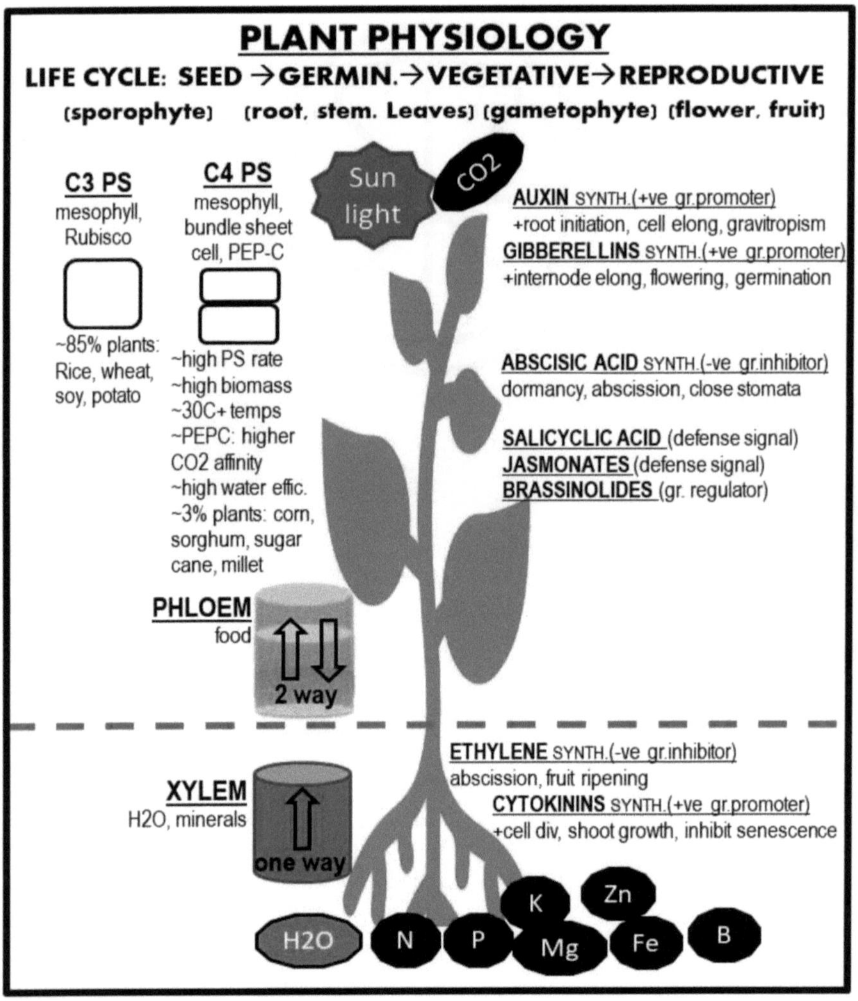

Fig. 4.1 Infographics of this chapter

4.3 Concepts of the Chapter

4.3.1 Transport

Plasma membrane is a selective barrier which allows specific molecules to pass through fast, slow, or blocked to control the cell traffic for normal functioning of the cell. The ability to pass membranes may be determined by the features such as size (small), hydrophobic nature, and charge (nonpolar or neutral).

Part 1: Passive Transport: This is a natural movement with concentration gradient and requires no energy, e.g., diffusion (simple), facilitated diffusion (with transporter proteins) (Fig. 4.2).

Part 2: Active Transport: This is an upstream movement and requires energy (ATP), e.g., pumps (Na/K), sugar transporters (from source to sink), nitrate transporters (Fig. 4.2).

For more on this topic, see Schroeder et al. (2013), Eichhorn et al. (2012), Hacisalihoglu (2011, 2022), Reece et al. (2014), Taiz and Zeiger (2006), and many references therein.

4.3.2 Water Potential

Water Potential (psi (Ψw)) is water's potential energy in megapascals (MPa) unit. The highest Ψw is pure water (0), and water always moves toward more negative Ψw. Transpiration is how H_2O evaporates from stomata (from soil to air cycle).

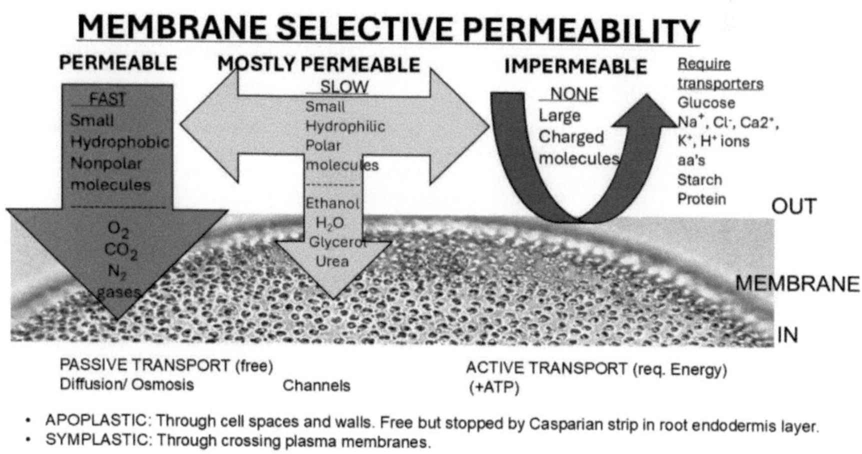

Fig. 4.2 Selective permeability of membranes and transport types

Other factors help water movements including cohesion, adhesion, and root pressure.

Ψw (water potential) = Ψp [solute potential] + Ψs [pressure potential]

(-) if tension; (+) if turgor (firm)

-100 MPa -20 MPa -7 MPa -0.6 MPa -0.3 Mpa 0

←air<-------dry seed <----leaf <-----xylem <-------soil <------- pure water

←.........water will move.......................................<

Ψw is always **negative** because energy is needed to be added to get H_2O out. For more on this topic, see Eichhorn et al. (2012), Hacisalihoglu (2011, 2022), Reece et al. (2014), Taiz and Zeiger (2006), and many references therein.

4.3.3 Plant Pigments Within Plastids

Plants and algae have plastid organelles and pigments within them. **Pigments** are usually assorted colors found in leaves, flowers, and fruits and play important roles in systems biology like photosynthesis and more. The three major plastids include chloroplasts, chromoplasts, and leucoplasts (no pigment) (Fig. 4.3).

For more on this topic, see Davies (2004) and many references therein.

GREEN pigments	YELLOW pigments	ORANGE pigments	RED/PURPLE pigments	BROWN pigments	WHITE pigments
Chlorophyll A (blue green) **Chlorophyll B** (green) **Chlorophyll C** (light green)	**Xanthophyll**	**Carotenes** (alpha and beta) **Lutein** (flavonoid)	**Lycopene** (red) **Anthocyanins** (purple) **Polyphenols**	**Tannins** **Phenolics**	**Allyl sulfides** **K-rich** (anti-diabetics)
Representative Species: *Green beans, kale, cabbage, lettuce, cucumber*	***Representative Species:*** *Banana, squash, pumpkin, avocado*	***Representative Species:*** *Sweet potatoes, papaya, pumpkin, peaches*	***Representative Species:*** *Tomato, strawberry, apple, grapes, blueberries*	***Representative Species:*** *Tea, coffee, cacao, legumes, berries*	***Representative Species:*** *Onion, garlic, leek*

Fig. 4.3 Six major plant pigments within plastids in the representative plant species

4.3.4 Photosynthesis Pathways (100+ Multistep Pathway)

Photosynthesis is considered the most important chemical reaction in biosphere because it releases oxygen for living organisms. Photosynthesis process has a total of 100+ steps. The two main stages of photosynthesis are light reactions and Calvin cycle (Fig. 4.4). The main organisms that carry out photosynthesis are **plants, algae, and cyanobacteria**. Photosynthesis takes place in chloroplast organelles that contain chlorophyll pigment. Photosynthesis uses sun energy, carbon dioxide (CO_2), and water to produce oxygen and sugar molecules (**G3P: glyceraldehyde 3P**) that store energy.

Other mechanisms on how photosynthesis occur are as follows:

- Cyanobacteria could be the first organism that used photosynthesis 3.5 Billion years ago, following the algae and plant evolution.
- Marine organisms (phytoplankton and seaweeds algae) produce over 50% of the O_2.
- It takes eight trees (100 kg O_2 production per tree) to provide enough O_2 for a person (usage of 740 kg O_2/year). Plants also absorb CO_2 (to convert to carbohydrates) as well as provide food, medicine, and other products.
- While **C3 photosynthesis** is considered the most common, **C4 photosynthesis** is adapted to warm temperatures, and **CAM photosynthesis** is adapted to hot desert conditions.
- C fixation is converting CO_2 (inorganic C) to organic compounds in mesophyll cells (C3 and C4).

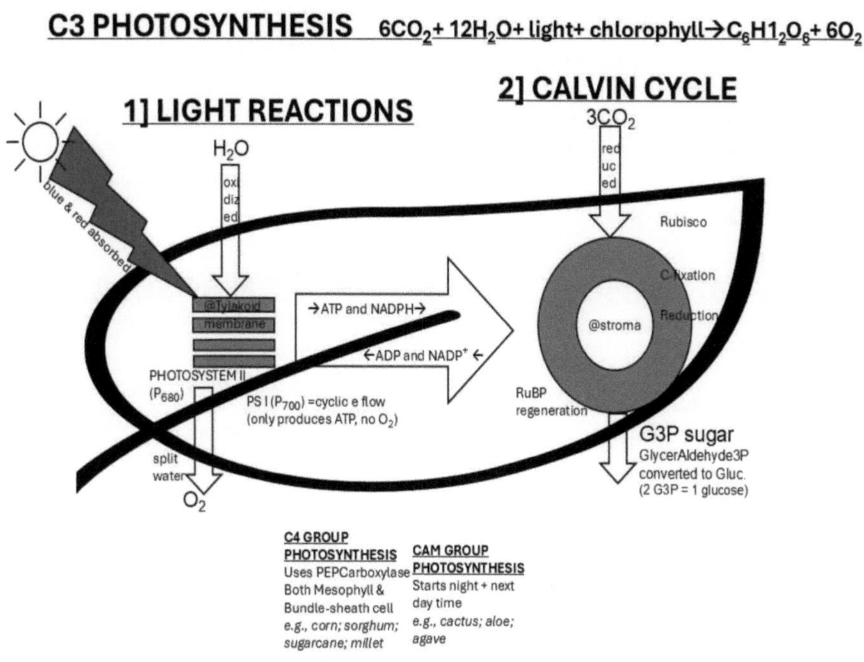

Fig. 4.4 Summary of C3 photosynthesis. C4 group uses PEPC as a CO_2-fixing enzyme in mesophyll cells and Rubisco as CO_2 assimilating enzyme in bundle sheath cells

- While C3 photosynthesis uses **Rubisco** (in mesophyll cells), C4 photosynthesis uses phosphoenolpyruvate carboxylase (PEPC) enzyme (in bundle sheath cells).
- C assimilation converts C to glucose (2 G3P equals to 1 glucose, Fig. 4.4).

For more on this topic, see Hacisalihoglu et al. (1999, 2001, 2003, 2007, 2008, 2018), Long et al. (2006), Marschner (1995), Schroeder et al. (2013), Taiz and Zeiger (2006), and many references therein.

4.3.5 Plant Hormones

Phytohormones or plant growth factors are small molecules with incredibly low concentrations and signal and coordinate various major physiological activities, including stress response, reproduction, formation of organs, and development. Some other plant processes using phytohormones include:

- **Senescence=** biological aging (e.g., leaf color changes in the fall season)
- **Abscission** = cutting or shedding off organs

While plants naturally produce their own hormones, there are synthetic hormones produced in the lab such as ethephon, which is synthetic ethylene in commercial fruit ripening usage. There are eight major phytohormones (Fig. 4.5).

In particular, auxin to cytokinin ratio is important in differentiation of root and shoot in callus cell cultures and promotes root growth (if AUX/CK high) or promotes shoot growth (if AUX/CK low).

For more on this topic, see Balick and Cox (1996), Eichhorn et al. (2012), Reece et al. (2014), FAO (2017, 2024), FAOSTAT (2023), Hacisalihoglu (2011, 2021, 2022), Harshberger (1896), Judd et al. (2015), Levetin and McMahon (2020), Prance et al. (2005), Ridolfi (2019), Schmidt and Cheng (2017), Simpson and Ogorzaly (2013), USDA (2023), USDA-NRCS (2000), Weber et al. (2024), Williams (2011), and many references therein.

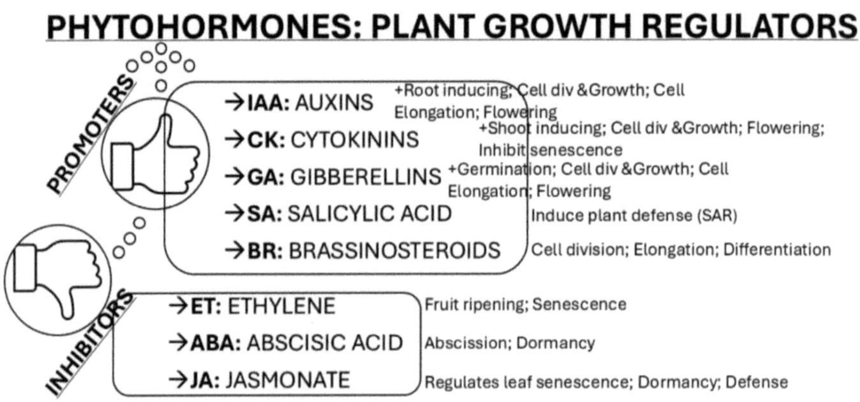

Fig. 4.5 Eight major plant hormones and their functions

4.4 Plant Profiles Bingo: How Many Can You Identify?

(P31)  *Crategus rhipidopphylla*	(P32)  *Cryptomeria japonica*
(P33)  *Cucurbita pepo*	(P34)  *Curcuma aromatica*
(P35) 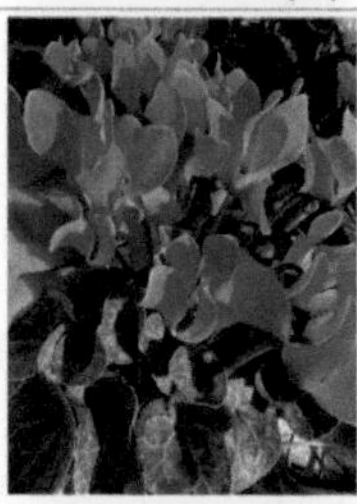 *Cyclamen persicum*	(P36)  *Datura metel*
(P37)  *Diandhus suprapink*	(P38)  *Dianthus caryophyllis*
(P39)  *Dieffenbachia seguine*	(P40)  *Digitalis purpurea*

4.5 Takeaway Questions

1. What is the role of auxin in plant growth and development?
2. What are the main stages of photosynthesis?
3. Which nutrient ions are transported in plants, and how are they absorbed?
4. What is the primary role of stomata in plants?

4.6 Now You Know the Answers

1. Auxin promotes cell elongation and regulates growth direction through phototropism and gravitropism, root development, and lateral root branching.
2. Light-dependent reactions occur in the thylakoid membranes of chloroplasts, and the Calvin cycle occurs in the stroma.
3. Essential nutrients like nitrogen, phosphorus, and potassium are absorbed by roots through active and passive transport mechanisms via transporter proteins.
4. Stomata regulate gas exchange, allowing CO_2 to enter for photosynthesis and releasing O_2 and water vapor via transpiration.

References

Balick MJ, Cox PA (1996) People, plants, and culture: the science of ethnobotany, New York

Davies KM (2004) An introduction to plant pigments in biology and commerce. In: Davies KM (ed) Plant pigments and their manipulation. Annual plant reviews, vol 14. Blackwell Publishing, Oxford, pp 1–22

Eichhorn SE et al (2012) Raven biology of plants, 7th edn. Macmillan

FAO (2017) The future of food and agriculture: trends and challenges. Food and Agriculture Organization of the United Nations, Rome, Italy

FAO (2024) Food and Agriculture Organization of the United Nations. Current market situation and medium-term outlook 1 August 2025. http://www.fao.org

FAOSTAT (2023) FAO statistical databases. Food and Agriculture Organization of the United Nations: Rome, Italy; https://www.fao.org/faostat

Hacisalihoglu G (2011) Plant biology concepts and laboratory manual. Kendall Hunt Publishing, Iowa

Hacisalihoglu, G. (2021). From growing to biology: plants 1st edn. Florida State University. 157

Hacisalihoglu G (2022) Level up college success: research, writing, and more key skills for STEM learners. ISBN13: 9798218066260. www.amazon.com/author/drhaci

Hacisalihoglu G, Taylor AG, Khan AA (1999) Embryo elongation and germination rates as sensitive indicators of lettuce seed quality. HortScience 34(7):1240–1243

Hacisalihoglu G, Hart JJ, Kochian LV (2001) High and low affinity Zn transport systems and their possible role in Zn efficiency in bread wheat. Plant Physiol 125:456–463

Hacisalihoglu G, Hart JJ, Cakmak I, Wang Y, Kochian LV (2003) Zinc efficiency is correlated with enhanced expression & Act. Of Cu/Zn-SOD & Carb. Anhydrase in wheat. Plant Physiol 131(2): 595–602

Hacisalihoglu G, Ji P, Longo LM, Olson S, Momol MT (2007) Bacterial wilt induced changes in nutrient distribution, biomass and the effect of Acibenzolar-S-methyl on bacterial wilt in tomato. Crop Prot 26:978–982

Hacisalihoglu G, Ji P, Olson S, Momol MT (2008) Effect of *Ralstonia solanaceaum* on mineral nutrients and IR temperatures in two tomato cultivars. J Plant Nutr 31(7):1221–1231

Hacisalihoglu G, Kantanka S, Miller N, Gustin JL, Settles AM (2018) Modulation of early maize seedling performance via priming under sub-optimal temperatures. PLoS ONE 13(11): e0206861. https://doi.org/10.1371/journal.pone.0206861

Harshberger JW (1896) The purposes of ethnobotany. Bot Gaz 21:146–154

Judd WS et al (2015) Plant systematics: a phylogenetic approach, 4th edn. Sinauer Associates, Sunderland

Levetin E, McMahon K (2020) Plants and society, 8th edn. McGraw Hill

Long SP et al (2006) Can improvement in photosynthesis increase crop yields? Plant Cell Environ 29:315–330

Marschner H (1995) Mineral nutrition of higher plants. Academic Press, London

Prance GT et al (2005) The cultural history of plants. Routledge, New York

Reece JB et al (2014) Campbell biology, 10th edn. Pearson, Boston

Ridolfi R (2019) SDGs for people, planet, and prosperity. http://www.openknowledge.fao.org/

Schmidt BM, Cheng DMK (2017) Ethnobotany: a phytochemical perspective. Wiley

Schroeder JI et al (2013) Using m. transporters to improve crops for sustainable food production. Nature 497:60–66

Simpson BB, Ogorzaly MC (2013) Plants in our world: economic botany, 4th edn. McGraw-Hill

Taiz L, Zeiger E (2006) Plant physiology, 4th edn. Sinauer Assoc. Inc.

USDA (2023) National Agriculture Statistics Service. https://www.nass.usda.gov/

USDA-NRCS (2000) The PLANTS database. National Resources Conservation Service. https://plants.usda.gov/

Weber C, Simnitt S, Wakefield H, Wechsler S, Swearingen B (2024) Fruit and tree nuts outlook: March 2024; United States Department of Agriculture: Washington, DC, USA. https://www.ers.usda.gov/publications/pub-details?pubid=108880

Williams ME (2011) Introduction to phytohormones teaching tools in plant biology: lecture notes. Plant Cell 22:1

Chapter 5
Genetics and Plant Breeding

Genetics and Plant Breeding

Contents

G. Hacisalihoglu, *Super Greens*, https://doi.org/10.1007/978-3-032-17047-7_5

5.1 Big Picture: Quick Read Chapter Infographic

5.2 Student Learning Outcomes (SLOs) of the Chapter

- Define polygenic inheritance and continuous variation.
- Summarize the major plant breeding techniques.
- Distinguish between various subfields in bioinformatics.
- Define codominance, allele, recessive, phenotype, and carrier.
- Identify monohybrid, dihybrid, and polygenic traits (Fig. 5.1).

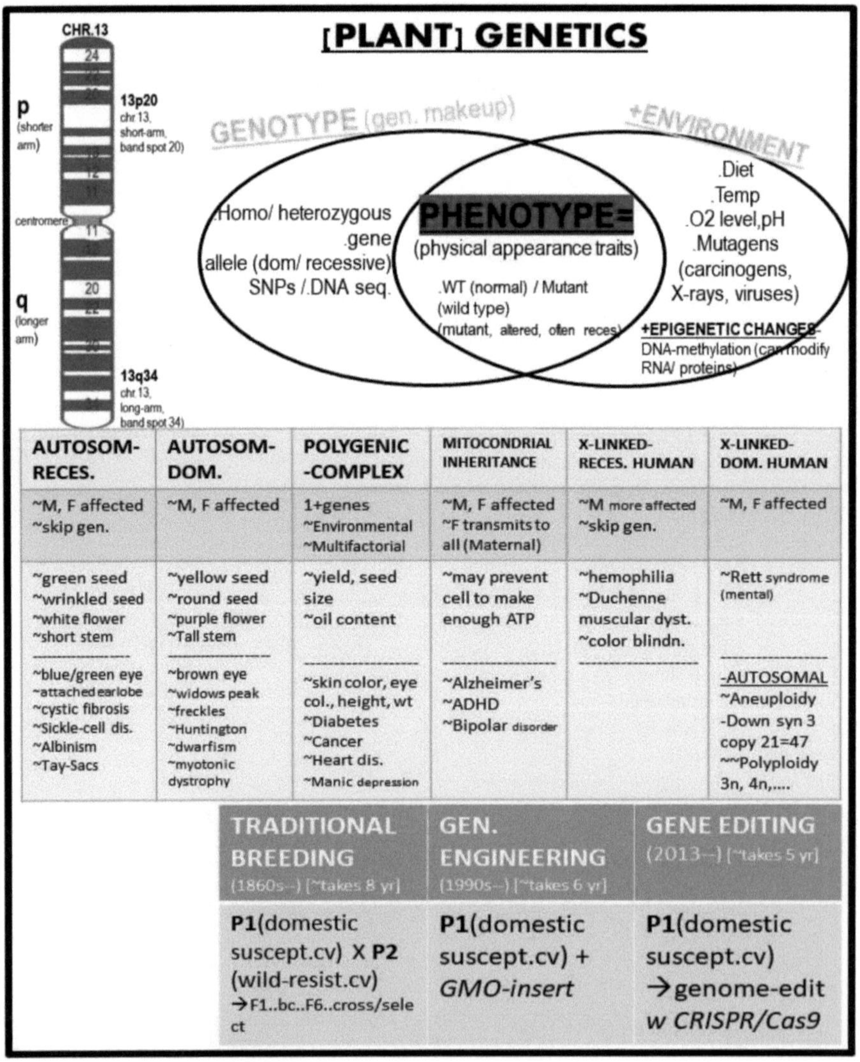

Fig. 5.1 Infographics of this chapter

5.3 Concepts of the Chapter

5.3.1 *Genes: Chunks of DNA Sequence*

Gene is a **DNA segment** on a chromosome with info to transcription (to mRNA) and translation (to protein). What are parts of a gene? While enhancers can increase transcription, terminator ends it. While messenger RNA (**mRNA**) carries info in DNA to ribosome, transfer RNA (tRNA) delivers amino acids (aa) to ribosome. It is replication if DNA-to-DNA, transcription if DNA-to-mRNA, translation if mRNA-to-protein, and reverse transcription if RNA-to-DNA. While *helicase* separates two DNA strands, *ligase* joins DNA strands back together. In the case of nucleotides, there are four of them that will appear in genes: adenine, cytosine, guanine, and thymine. Exon is the transcribed section of the gene that is found in mRNA (Fig. 5.2).

In the case of transcription, ***RNA polymerase II*** binds to promoter to synthesize template strand into mRNA. In the case of translation, ribosome binds to mRNA's AUG codon (start codon) and tRNA with the right amino acid (methionine for start) binds as well (initiation stage). Next, elongation adds more amino acids, and termination releases the final polypeptide (protein) (Fig. 5.2).

For more on this topic, see Acquaah (2012), AG Initiative (2000), Bailey-Serres et al. (2019), Eichhorn et al. (2012), Reece et al. (2014), and many references therein.

5.3.2 *Mendelian Genetics*

Classical genetics, study of genes passed to offspring, was first studied and published by Gregor Mendel in **1866**. Mendel carried out his experiment with

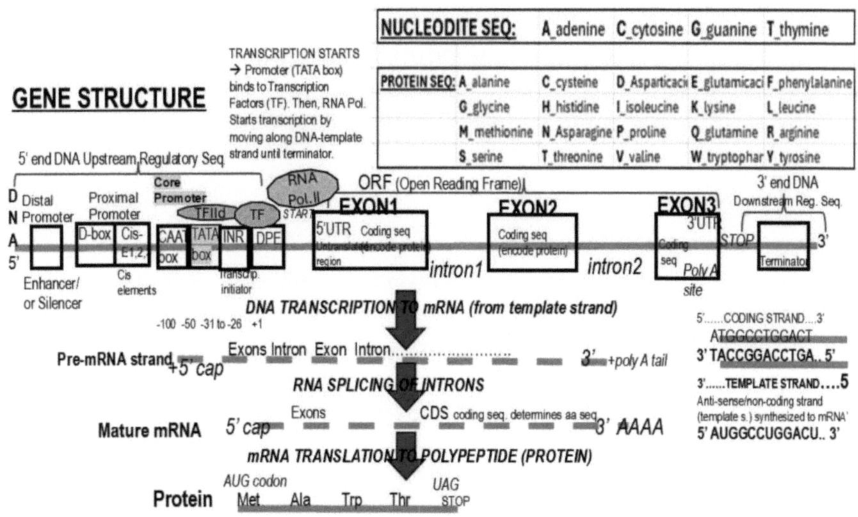

Fig. 5.2 A summary of gene structure, transcription, and translation processes

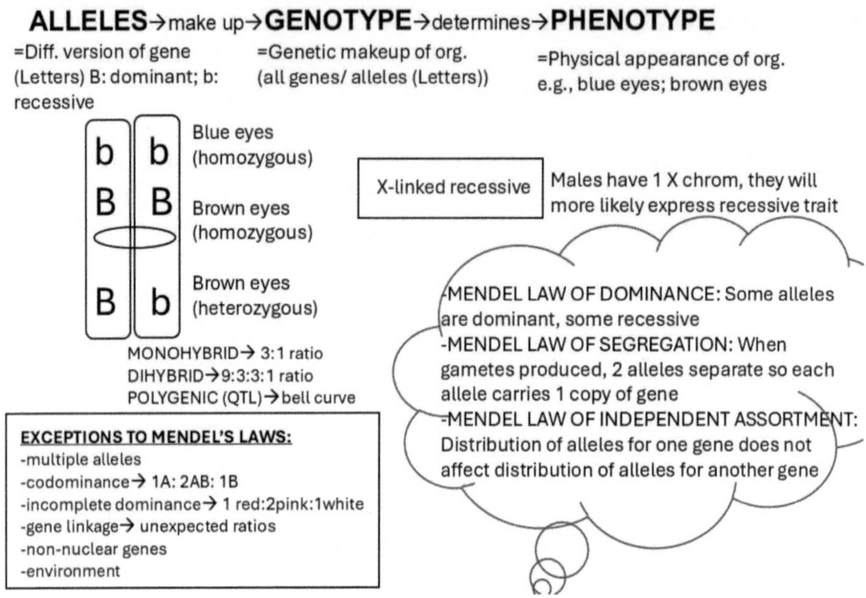

Fig. 5.3 Mendelian genetics basics

model plant peas (*Pisum sativum*) with great genetic variation and rapid life cycle. Some of the inheritance patterns include:

- **Recessive** traits show if two copies of alleles are inherited. If one copy is inherited, it is called "carrier (heterozygous)." As an example, blue eyes are recessive to brown eyes, but some brown eye individuals would be carriers, and therefore blue eyes continue to show (8% of the population) since the **first blue eye** person from Black Sea 7000 years ago (Fig. 5.3).

For more on this topic, see Beyer (2010), Beracochea et al. (2023), and many references therein.

5.3.3 *Polygenes*

Polygenes are caused by combo contribution of a number of genes to influence a single trait. Polygenes produce continuous phenotypes variation. In the case of humans, skin color, height, and weight are controlled by polygenes. In the case of plants, seed color, seed size, and yield are controlled by polygenes (Fig. 5.4).

For more on this topic, see Blair et al. (2006), Borlaug (1983), Ceballos et al. (2004), Gross et al. (2003), Grusak and DellaPenna (1999), Hacisalihoglu (2021), Hacisalihoglu et al. (2008, 2018), Jankowicz-Cieslak and Till (2015), Jones and Dangl (2006), Potrykus (2001), Renard and Tilman (2019), Saltzman et al. (2017), Sen (2004), Schopfer et al. (1999), Tanaka and Fujiwara (2008), and many references therein.

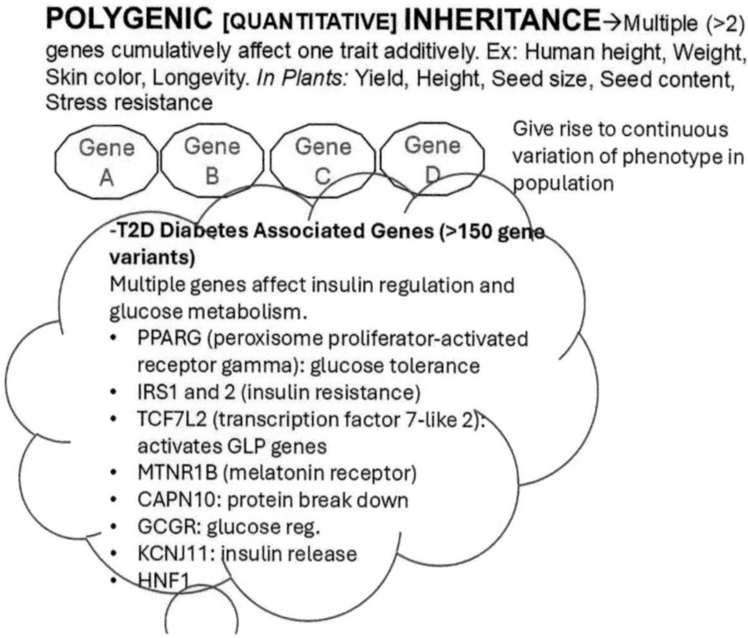

Fig. 5.4 Polygenic inheritance and quantitative traits, e.g., type 2 diabetes in humans

5.3.4 Plant Breeding

Breeding crops aims to improve certain more desirable traits in plants. Generations in plants include **parental (P), filial first, second, and third (F1, F2, F3)**. If you cross two pure lines, you get "F1 Hybrid," which are homogenous and more vigorous than parents (heterosis or hybrid vigor with improved traits). Some **F1 hybrids** with higher yield or greater resistance to diseases include corn hybrids. However, if you grow seeds from the ears of F1 hybrids, you will get F2 generation (not same as F1).

While **point mutation** is a sudden variation of gene (inherited), it can potentially cause changes in expressed gene and protein. While the **pedigree** is a record of individual's ancestry, it can show inheritance patterns and their possible tracking. Polyploids have more than two sets of chromosomes (e.g., triploid, tetraploid). Genotype identification can be carried out using the techniques as follows: polymerase chain reaction (PCR), **DNA sequencing**, as well as test-cross (determine genotype of unknown parent). While variety occurs naturally, cultivars are specifically man-made varieties (e.g., granny smith apple). Some of the plant breeding techniques include classical (selective) breeding, hybrid creation, genetic engineering, molecular breeding, and genome editing (Fig. 5.5). While transgenic GMO crops use foreign gene(s), gene editing uses plant's own DNA to improve favorable traits such as disease resistance or crop yields.

For more on this topic, see Allard (1999)) and many references therein.

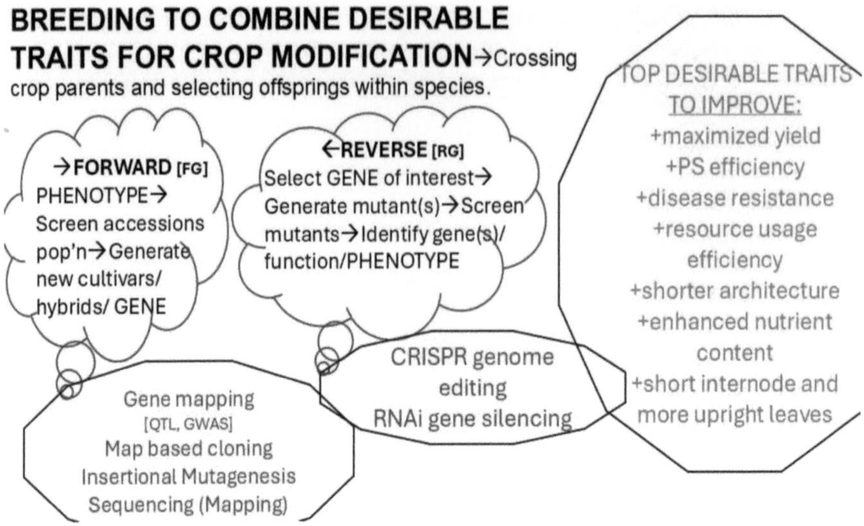

Fig. 5.5 Plant breeding and crop improvement for featuring favorable quality traits, including greater shelf life, breeding cycles, yield, nutrient content, and withstand pests and diseases

5.3.5 Bioinformatics: Analyze, Annotate, and Reformat the Sequence Data

Bioinformatics uses computer programming and statistics to interpret sequences (DNA, protein). Genome sequencing is a complete nucleotide sequence of a plant's DNA. In the case of gene annotation, scientists identify and describe gene functions in a sequence of a genome (Fig. 5.6). Some of the genetics and bioinformatics techniques are as follows:

- Single-nucleotide polymorphism (SNP) is a single base pair difference in a sequence, e.g., AGCTA vs AGCTT.
- In the case of well-annotated genomes, the top five plant species are *Arabidopsis*, rice, corn, wheat, barley.
- Bioinformatics: Analyze, annotate, and reformat the sequence data for databases.

Some of the bioinformatics milestone happenings are as follows:

- ***Arabidopsis*** Genome sequence completion (2000) by the *Arabidopsis* Genome Initiative (25,000 genes)
- **Human** Genome sequence completion (2004) by the International Human Genome Sequencing Consortium
- **Soybean** (var. Williams 82) Genome sequence completion (2010)
- **Maize** Genome sequence completion (2021) by Hufford et al.
- **Rice** Genome sequence completion (2018) by Zhao et al.
- **Barley** Genome sequence completion (2012) by the International Barley Genome Sequencing Consortium

BIOINFORMATICS→Computational in-silico biology.
SEQUENCE→STRUCTURE→FUNCTION CHARACTERIZATION

RAW GENOME SEQUENCE ATCGTACG [Adenine, Thymine, Cytosine, Guanine]	TOOL/ ALGORITHMS	BIOINFORMATICS ANALYSIS
i) SEQUENCE ANALYSIS / SEQ. ALIGNMENT	-BLAST -CLUSTALW	-Similarity search DNA seqs relevant to your protein -Multiple protein seq. alignment
ii) GENE PREDICTION & PROMOTER PREDICTION ALGORITHMS	-HOMOLOGY -AB-INITIO	-compare seq. to known genes in similar organisms -based on stat. patterns from given sequence alone
iii) FUNCTIONAL GENE ANNOTATION MANUAL CURATION STRUCTURE ANALYSIS	-ML MODELS -GRAINGENES -JBROWSE -APOLLO -MAKER -GENSAS	-ID & describe: dna/prot seq, seq corrections, functions, abundance, location, homologues, pathways -Metabolic pathway modeling, Protein interaction prediction, Protein subcel. localization prediction -structure prediction of nuc. acids, protein -structure comparison
HIGHLY CURATED DATABASE	VISUALIZATION	-Make them publicly available

Fig. 5.6 Bioinformatics subfields and applications including genome annotation and curation

- **Bread wheat (*Triticum aestivum*) variety Chinese Spring (CS)** by Genome sequence completion (2018) by the International Wheat Genome Sequencing Consortium (IWGSC). Six sets of chromosomes, 42 chromosomes ($2n = 6\times = 42$), consisting of three 2n genomes (A, B, or D genome); 96,000 genes

What about **annotation**? Genome annotation reads and interprets DNA sequences to identify and describe gene functions. Functional annotation specifically focuses on proteins and RNA regulation as well as the pathways they participate in. Genome browsers can help displaying and analyzing genomic data. Examples of some genome browsers are GrainGenes (https://wheat.pw.usda.gov/GG3/genome_browser), Ensembl (https://plants.ensembl.org/index.html), and NCBI (https://www.ncbi.nlm.nih.gov/gdv/).

What about **curation**? Genome curation reviews and organizes data on genes and related info for accuracy as well as gene identification and database creation (Fig. 5.6).

For more on this topic, see Blake et al. (2019), Schmutz et al. (2010), Strable and Scanlon (2009), The Arabidopsis Genome Initiative (2000), The International Barley Genome Sequencing Consortium (2012), The International Human Genome Sequencing Consortium (2004), and many references therein.

For more on this topic, see Cousins (2022), Zhu et al. (2021), Zhao et al. (2018), Welch and Graham (2004), Balick and Cox (1996), Eichhorn et al. (2012), Reece et al. (2014), FAO (2017, 2024), FAOSTAT (2023), Hacisalihoglu (2011, 2021, 2022), Harshberger (1896), Hufford et al. (2019), Judd et al. (2015), Levetin and McMahon (2020), Prance et al. (2005), Ridolfi (2019), Schmidt and Cheng (2017), Simpson and Ogorzaly (2013), USDA (2023), USDA-NRCS (2000), Weber et al. (2024), Kantoglu (2014), Guner and Wehner (2004), and many references therein.

5.4 Plant Profiles Bingo: How Many Can You Identify?

(P41)  *Dionaea muscipula*	(P42) 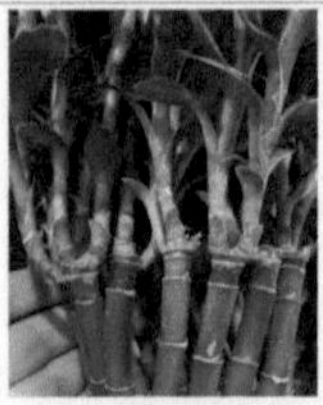 *Dracaena sanderiana*
(P43)  *Drosera capensis*	(P44)  *Drosera sundews*
(P45)  *Dryopteris erythrosora*	(P46)  *Echeveria elegans*
(P47)  *Echinocereus striglochidatus*	(P48)  *Equisetum telmateia*
(P49)  *Euphorbia pulcerrima*	(P50)  *Ferocactus emoryi*

5.5 Takeaway Questions

1. What is the importance of genetic diversity in plants?
2. Why did Mendel use pea plants in his experiments?
3. How are quantitative traits different from qualitative traits?
4. What is the primary purpose of genome annotation in bioinformatics?
5. What is a sequence alignment in bioinformatics?

5.6 Now You Know the Answers

1. It ensures survival of plant species by providing resilience to environmental changes, diseases, and pests, among others.
2. Because they have easily observable traits, short generation time, and self/cross-pollination for breeding.
3. Quantitative traits exhibit continuous variation (e.g., yield), while qualitative traits show distinct categories (e.g., flower color: purple or white).
4. To identify genes and regulatory elements.
5. A method of arranging sequences to identify regions of similarity to find functional, structural, or evolutionary relationships.

References

Acquaah G (2012) Principles of plant genetics and breeding. Wiley

AG Initiative (2000) Analysis of the genome sequence of the flowering plant *Arabidopsis thaliana*. Nature 408:796–815

Allard, R. W. (1999) Principles of plant breeding. 2nd edn Wiley 264

Bailey-Serres J et al (2019) Genetic strategies for improving crop yields. Nature 575:109–118

Balick MJ, Cox PA (1996) People, plants, and culture: the science of ethnobotany, New York

Beracochea V et al (2023) CRISPR/Cas9-mediated knockout of SPL13 radically increases lettuce yield. Plant Cell Rep 42:645–647

Beyer P (2010) Golden rice and 'golden' crops for human nutrition. New Biotechnol 27:478–481

Blair MW et al (2006) QTL analysis of yield traits in an advanced backcross population derived from a cultivated Andean × wild common bean (*Phaseolus vulgaris* L.) cross. Theor Appl Genet 112:1149–1163

Blake VC, Sen TZ et al (2019) GrainGenes: centralized small grain resources and digital platform for geneticists and breeders. Database:baz065. https://doi.org/10.1093/database/baz065

Borlaug NE (1983) Contributions of conventional plant breeding to food production. Science 219: 689–693

Ceballos H et al (2004) Cassava breeding: opportunities and challenges. Plant Mol Biol 56:503–516

Cousins P (2022) Grapevine breeding: progress, innovation and opportunity. Julius-Kuhn-Arch

Eichhorn SE et al (2012) Raven biology of plants, 7nd edn. Macmillan

FAO (2017) The future of food and agriculture: trends and challenges. Food and Agriculture Organization of the United Nations, Rome, Italy

FAO (2024) Food and Agriculture Organization of the United Nations. Current market situation and medium-term outlook 1 August 2025. http://www.fao.org

FAOSTAT (2023) FAO statistical databases. Food and Agriculture Organization of the United Nations: Rome, Italy; https://www.fao.org/faostat

Gross J et al (2003) Iron homeostasis related genes in rice. Gen Mol Biol 26:477–497

Grusak MA, DellaPenna D (1999) Improving the nutrient composition of plants to enhance human nutrition and health. Ann Rev Plant Physiol Plant Mol Biol 50:133–161

Guner N, Wehner TC (2004) The genes of watermelon. J Am Soc Hortic Sci 6:1175–1182

Hacisalihoglu G (2011) Plant biology concepts and laboratory manual. Kendall Hunt Publishing, Iowa

Hacisalihoglu, G. (2021). From growing to biology: plants 1st edn. Florida State University. 157

Hacisalihoglu G (2022) Level up college success: research, writing, and more key skills for STEM learners. ISBN13: 9798218066260. www.amazon.com/author/drhaci

Hacisalihoglu G, Hilgert U, Nash EB, Micklos DA (2008) An innovative Plant Genomics & Gene Annotation Program for HS, CC & University Faculty. Cell Biology Education-LSE 7(3): 310–316

Hacisalihoglu G, Burton AL, Eker S, Ozturk L, Cakmak I, Burkey KO, Orf J, Settles M (2018) Quantitative trait loci associated with soy seed weight and composition under different P levels. J Integr Plant Biol 60(3):232–241. https://doi.org/10.1111/jipb.12612

Harshberger JW (1896) The purposes of ethnobotany. Bot Gaz 21:146–154

Hufford MB et al (2019) Crop biodiversity: an unfinished magnum opus of nature. Annu Rev Plant Biol 70:727–751

Jankowicz-Cieslak J, Till BJ (2015) Forward and reverse genetics in crop breeding. In: Advances in plant breeding strategies: breeding, biotechnology and molecular tools, pp 215–240

Jones JDG, Dangl JL (2006) The plant immune system. Nature 444:323–329

Judd WS et al (2015) Plant systematics: a phylogenetic approach, 4th edn. Sinauer Associates, Sunderland

Kantoglu KY et al (2014) Vegetable crops breeding by induced mutation and a practical case study of *Capsicum annuum* L. In: Tomlekova NB (eds) Mutagenesis: exploring genetic diversity in crops. Wageningen Publishers, pp 41–56

Levetin E, McMahon K (2020) Plants and society, 8th edn. McGraw Hill

Potrykus I (2001) Golden rice and beyond. Plant Physiol 125:1157–1161

Prance GT et al (2005) The cultural history of plants. Routledge, New York

Reece JB et al (2014) Campbell Biology, 10th edn. Pearson, Boston

Renard D, Tilman D (2019) National food production stabilized by crop diversity. Nature 571:257–260

Ridolfi R (2019) SDGs for people, planet, and prosperity. http://www.openknowledge.fao.org/

Saltzman A et al (2017) Availability, production, and consumption of crops biofortified by plant breeding: current evidence and future potential. Ann N Y Acad Sci 1390:104–114

Schmidt BM, Cheng DMK (2017) Ethnobotany: a phytochemical perspective. Wiley

Schmutz J et al (2010) Genome sequence of the paleopolyploid soybean. Nature 463(7278): 178–183

Schopfer CR et al (1999) The male determinant of self-incompatibility in brassica. Science 286: 1697–1700

Sen TZ (2004) Predicting binding sites of hydrolase-inhibitor complexes by combining several methods. BMC Bioinformatics 5:1–11

Simpson BB, Ogorzaly MC (2013) Plants in our world: economic botany, 4th edn. McGraw-Hill

Strable J, Scanlon MJ (2009) Maize (*Zea mays*): a model organism for basic and applied research in plant biology. Cold Spring Harb Protoc 10:pdb. emo132

Tanaka M, Fujiwara T (2008) Physiological roles and transport mechanisms of boron: perspectives from plants. Eur J Phys 456:671–677

The Arabidopsis Genome Initiative (2000) Analysis of the genome sequence of the flowering plant *Arabidopsis thaliana*. Nature 408:796–815

The International Barley Genome Sequencing Consortium (2012) A physical, genetic and functional sequence assembly of the barley genome. Nature 491:711–716. https://doi.org/10.1038/nature11543

The International Human Genome Sequencing Consortium (2004) Finishing the euchromatic sequence of the human genome. Nature 431:931–945. https://doi.org/10.1038/nature03001

USDA (2023) National Agriculture Statistics Service. https://www.nass.usda.gov/

USDA-NRCS (2000) The PLANTS Database. National Resources Conservation Service. https://plants.usda.gov/

Weber C, Simnitt S, Wakefield H, Wechsler S, Swearingen B (2024) Fruit and tree nuts outlook: March 2024; United States Department of Agriculture: Washington, DC, USA. https://www.ers.usda.gov/publications/pub-details?pubid=108880

Welch RM, Graham RD (2004) Breeding for micronutrients in staple food crops from a human nutrition perspective. J Exp Bot 55:353–364

Zhao Q et al (2018) Pan-genome analysis highlights the extent of genomic variation in cultivated and wild rice. Nat Genet 50:278–284

Zhu T et al (2021) Optical maps refine the bread wheat Triticum aestivum cv. Chinese spring genome assembly. Plant J 107:303–314

Chapter 6
Planting and Horticulture

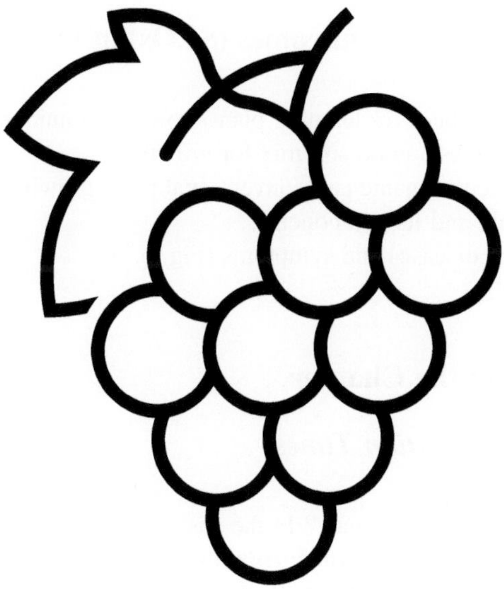

Planting and Horticulture

Contents

G. Hacisalihoglu, *Super Greens*, https://doi.org/10.1007/978-3-032-17047-7_6

6.1 Big Picture: Quick Read Chapter Infographic

6.2 Student Learning Outcomes (SLOs) of the Chapter

- Apply the best practices for learning plant growing techniques.
- Describe what makes a good soil mix for growth.
- Distinguish between organic and conventional plant growing.
- Define plant yield and its components.
- Understand plant diseases and symptoms (Fig. 6.1).

6.3 Concepts of the Chapter

6.3.1 Seed Germination Times

Consider germination vs dormancy? In the case of **germination**, seed sprouts and starts plant growth. In the case of **dormancy**, seeds cannot germinate even after exposure to proper growth conditions (Fig. 6.2).

Seeds need to be selected for seed type, including hybrid (uniform, high yield but used one time only) or heirloom (old varieties and can be saved for generations). While in epigeal germination, cotyledons rise above ground (beans, pumpkin, tomato), in hypogeal germination, they stay below ground (corn, wheat, rice).

While germination time is **faster for lettuce**, spinach, and some herbs (24 h), it is very slow germination for palms, cycads, yews, lavender, asparagus, okra, and parsley (up to 30 days) (Fig. 6.2).

For more on this topic, see Bewley et al. (2013), Bradford and Nonogaki (2007), and many references therein.

Part A. Know Your Hardiness Zone (min. Cold Temperatures) and Growing Season Length: The growing **season** is a period from spring's last frost to fall's first frost. While Alaska has the shortest growing season (105 days), zone 10 has the

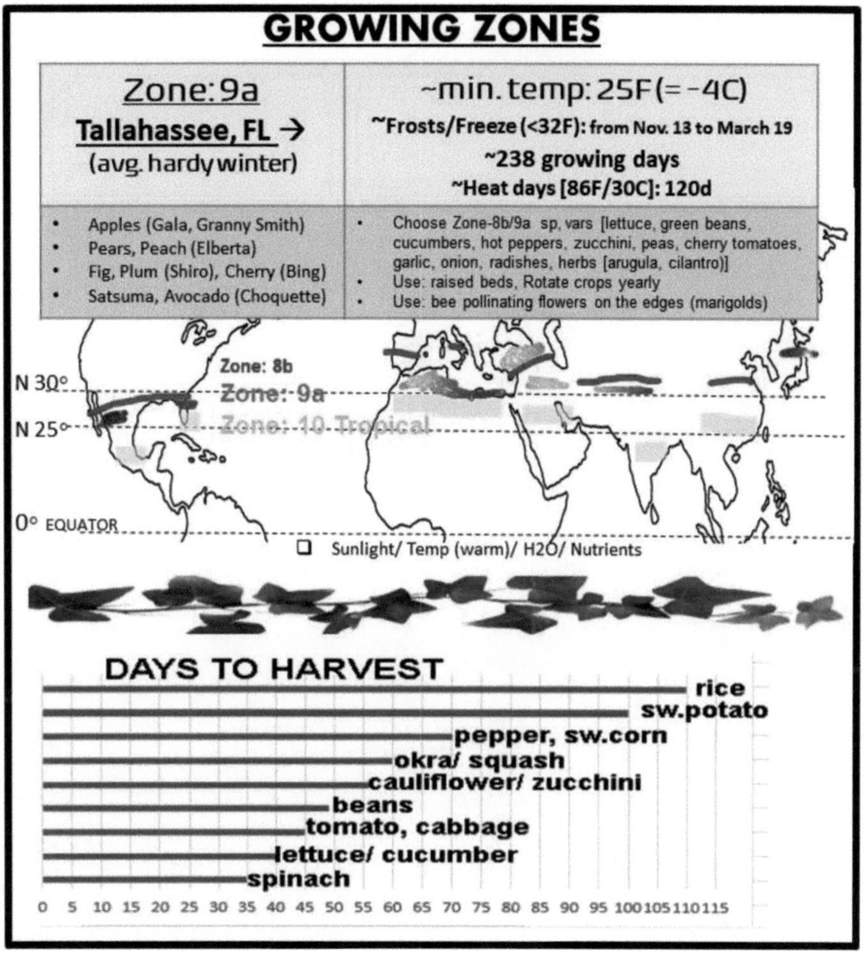

Fig. 6.1 Infographics of this chapter

longest growing season (South FL, Southern CA, and Hawaii). The closer to the equator, the longer the growing season. For example, some of the growing season features of Tallahassee, Florida, are as follows: 30 N (latitude) 84 W (longitude), growing season of 240 days (March 19 to November 15). USDA plant hardiness zone maps can help plants thrive in their specific region and climate. For example, Tallahassee, Florida, is in **zone 9a** (20F–25F), which allows the cultivation of subtropical plant species. This is subtropical zone (23 N to 356 N), including Florida, California, Spain, Italy, Greece, France, Japan, and Türkiye.

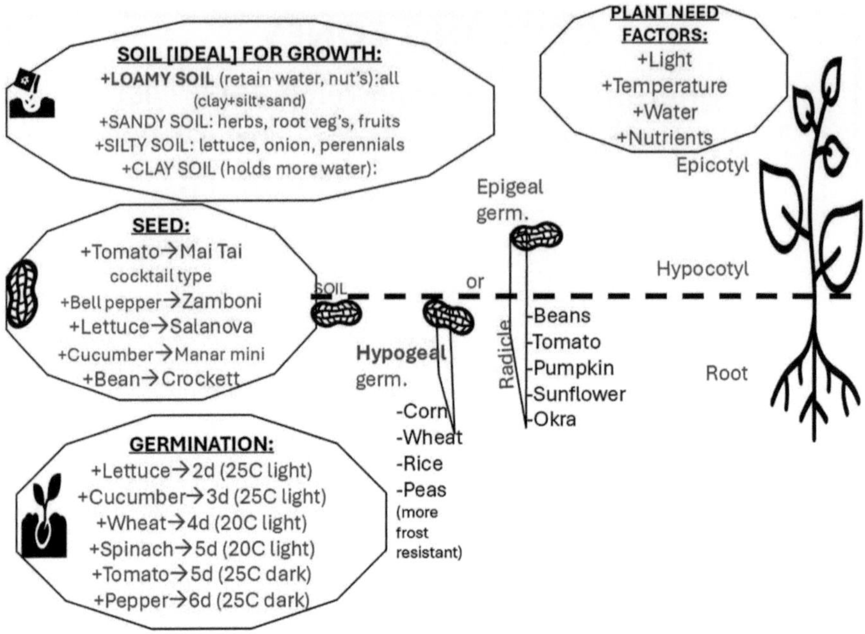

Fig. 6.2 Seeds, germination, and growth factors plants need

6.3.2 *Plant Nutrition and Deficiencies Symptoms*

There are **17 essential nutrients**: nitrogen (N), phosphorus (P), potassium (K), calcium (Ca), magnesium (Mg), sulfur (S), chloride (Cl), iron (Fe), boron (B), manganese (Mn), zinc (Zn), copper (Cu), molybdenum (Mo), and nickel (Ni) (plus C, H, O non-mineral elements). While macronutrients are needed in large amounts, trace elements are needed in small amounts. In the case of beneficial elements, cobalt, silicon, and sodium could be listed. There are natural or synthetic fertilizers or plant food with essential nutrients that growers usually add to their soil to promote plant growth. One key feature on plant fertilizers is NPK ratio that indicates nitrogen, phosphorus, and potassium ratios. For example, 10-5-8 label shows that it contains 10% N, 5% P, and 8% K (Fig. 6.3).

Most plants grow in soil where they obtain their water and essential nutrients. Although they vary, soil depth could be 1–3 meters. In these soil profiles, there are five different profiles, including O, A, B, C, and R horizons of soil (Fig. 6.3).

The key soil types of numerous particles are as follows: **SAND (2 mm)** . . . **SILT (0.05 mm)** . . . **CLAY (0.002 mm)** (Fig. 6.4).

Clay (0.002 mm chalky, poor drainage), loamy (highly fertile soil), peaty (wet climate soil), silty (0.05 mm slow drain), and sandy (2 mm quick drainage). Loamy soils contain a mixture of clay (20%), sand (40%), and silt (40%) as well as excellent capacities of nutrients, water, and drainage. Nutrient-rich high organic matter fertile

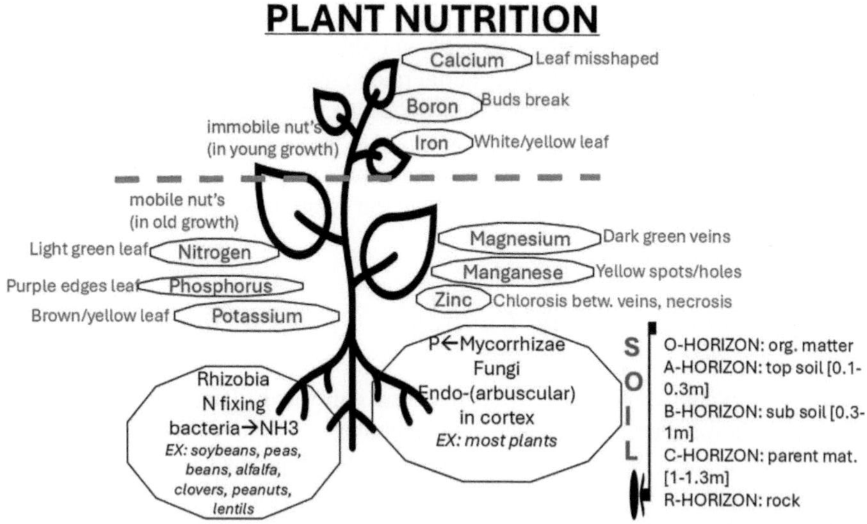

Fig. 6.3 Plant nutritional biology and soil horizons. Macro- and micronutrients are typically absorbed from the soil via plant roots for plant growth and development. Mineral nutrients (inorganic) could be immobile (B, Cu, Ca, Fe, S) and therefore may have lower translocation during their deficiencies. Major soil types (horizons) from top to down are O, A, B, C, and R; high N-needed plants: bok choy, lettuce, spinach, kale, cabbage, beets, and broccoli. High P-needed plants: wheat, onions, potato, and carrot. High K-needed plants: tomato, pepper, eggplant, and flowers [roses, lilies, dahlias]

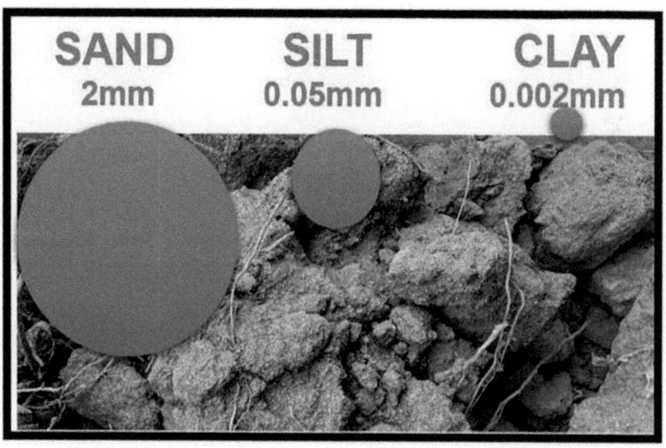

Fig. 6.4 Different soil types and their particle sizes: sand, silt, and clay. (Image: G. Hacisalihoglu, PhD)

soils can be found in the USA (soft **mollisol soils** in Iowa, Illinois, Nebraska, Indiana, and Ohio), Ukraine (chernozem mollisol soils), north China, Argentina (Pampas), and Mediterranean (alluvial soil) and Aegean region of Türkiye.

While **mycorrhizae** are beneficial fungi that provide **P** to plants, rhizobia are beneficial bacteria that provide N to legume plants (beans, peas, soybeans). Therefore, legumes are subsequently protein-rich plant species due to increased N access levels.

How would you know if a plant is deficient in a nutrient? Nutrient deficiency symptoms, including chlorosis, necrosis, and short internodes, allow growers to identify missing or low nutrients. Furthermore, chemically immobile elements such as Fe, Ca, and B always show their deficiency symptoms in young growth or upper parts of the plants (Fig. 6.3). For example, Fe is an immobile element, and younger leaves of roses will show the symptoms. On the other hand, N is a mobile element, and older leaves of beans will show the symptoms.

6.3.3 *Plant Growing and Cultivation*

While wild plants could grow by themselves, cultivated plants are cultivated by growers in horticulture (art, science, and technology of plant cultivation). Potatoes were first cultivated in Peru, where their center of origin is located. In the case of hydroponics, plants grow without soil, including water (Fig. 6.5).

Part A. Low or High pH Preferring Plants (<5 Very Acidic, >7 Very Alkaline): While conventional (industrial) horticulture typically uses GMO crops, synthetic fertilizers, and pesticides, organic growing uses more natural resources. Soil pH is a feature of how acidic or basic the soil is. While most plants grow at a pH

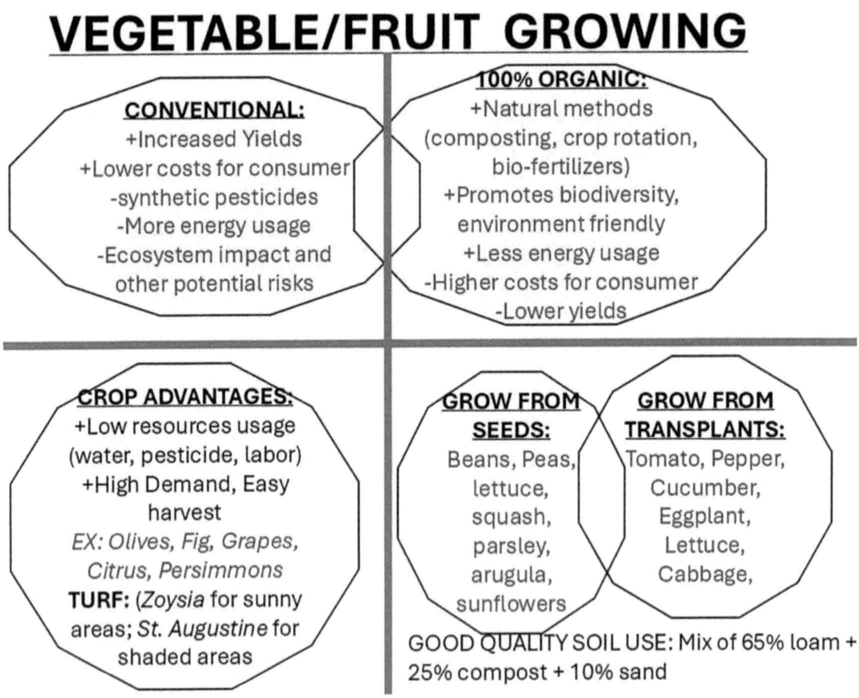

Fig. 6.5 Industrial (conventional) and organic horticulture in terms of crops, varieties, and soils

of 6–7, acid loving plants grow at a pH of 5–7 (tea, azalea, camellia, citrus, blueberry, heather, magnolia). It is very acidic at pH 5 and below. The **varying colors** of hydrangeas are due to soil pH levels (blue: pH 5; purple pH 6; and pink pH 7). While soil pH can be increased with adding lime, it can be decreased with adding sulfur). Representing the most fertile soils, loamy soils are considered organic because they contain substantial amount of humus, sand, silt, and clay materials mix. Lastly, variegated plants contain alternating multicolor foliage (Fig. 6.5).

6.3.4 Organic Certification

In the case of the **100% organic certified label**, plants are grown and have undergone USDA certification. Additionally, farms require a 3-year transition that will only be farmed organically continuously. As an organic fertilizer, organic farms use compost, which is a mix of soil and decaying organic material. Some of the most common organic growing features include as follows (Fig. 6.6; Table 6.1):

- **Organic Certification:** Inspected and verified with USDA Organic Seal
- **Cover Crops:** Cover between rows and control weeds (e.g., clover)
- **Crop Rotation:** Rotating crop locations each season (roots: onion, beets, carrot, celery), tomato, potato, legumes (beans, peas), and Brassicas (cabbage, kale, cauliflower). High N-needed plants: bok choy, lettuce, spinach, kale, cabbage, beets, broccoli. High P-needed Plants: wheat, onions, potato, carrot. High K-needed Plants: tomato, pepper, eggplant, flowers [roses, lilies, dahlias]

Fig. 6.6 Organic vegetable growing is based on using sustainable techniques, healthy soils, no synthetic fertilizers, no GMO seeds, and no pesticides. While beans, peas, spinach are planted as direct sowing, cabbage, broccoli, and cauliflower are planted as transplanting seedlings. (Image: G. Hacisalihoglu, PhD)

Table 6.1 Most popular crop species, varieties, cultivars, and yield estimates from a 10 m^2 garden

Crop species	Fav. cultivars	Yield (kg)
		(3 m × 3m)- > 10 m^2
Tomato	'Sungold' (cherry)	113
	'Alston Everlasting'	
	'Juliet'	
	'Tiny Tim'	
	'Micro Tom'	
	'Dwarf Eagle Smiley'	
Pepper	'Shishito'	54
	'Jimmy Nardello'	
	'Olympus'	
Beans	'Dragon's Tongue'	45
Peas	'Sugar Snap'	45
Kale	'Dinosaur'	54
Lettuce	'Little Gem'	144
Cucumber	'Blue Lake'	162
Onions	'Walla Walla'	135
Potatoes	'Yukon Gem'	135
Corn	'Silver Queen'	25
Wheat	'Hard Winter'	9
Leaks	'King Richard'	225
Cabbage	'Tendersweet'	135
Carrots	'Red Cored'	158
Broccoli	'Green Magic'	34
Radish	'French Breakfast'	68

- **Sustainable Ag:** Maintain nature and mitigate negative impacts
- **Regenerative Ag:** Restore and improve nature

6.3.5 *Chemical Fertilizers*

Chemical fertilizers (**synthetic fertilizers**) have the ability to improve soil structure and plant nutrition. Plants need 17 essential nutrients for growth and health. Plants are in need of nitrogen (N) the most. After N, phosphorus (P) and potassium (K) are needed the most (Fig. 6.7). **NPK** number is the standard fertilizer combination for % N, % P, and % K by volume. For example, NPK of 20: 20: 20 label represents equal percentages (20%) of N, P, and K by volume in the fertilizer bag.

For more on this topic, see Marschner (1995), Hacisalihoglu (2021), Galuzzi et al. (2010), Dragland et al. (2003), Black and Bewley (2000), Arnon and Hoagland (1940), and many references therein.

Fig. 6.7 Chemical fertilizers are less costly and more likely to work quicker to improve plant growth and health. (Image: G. Hacisalihoglu, PhD)

6.3.6 Biofertilizers

Biofertilizers, eco-friendly beneficial organisms-based fertilizers, can enhance soil heath and plant productivity as an alternative to chemical fertilizers. Biofertilizers may help with nitrogen fixing, phosphorus solubilizing, or other nutrient supplying to improve soil and plant nutrients (Table 6.2).

For more on this topic, see Ibanez et al. (2023) and many references therein.

6.3.7 Plant Diseases (Pathogen + Host + Environment)

Pathogens are infectious biotic agents such as bacteria, viruses, fungi, and nematodes that can cause plant diseases. A disease triangle can develop in the presence of a pathogen agent and conducive environment (rain, temperature, soil).

The most common plant disease **symptoms** are wilts, vascular wilts, blights, damping off, root rots, galls, and cankers (Fig. 6.8). While chlorosis is yellowing, canker is a dead area, stunting is no growth, wilt is drooping water loss, and blight is general discoloration in leaves.

Fungi cause 85% of plant diseases and are typically spread by contaminated seeds, wind, rain, insects, and birds, among others. Most well-known fungal disease is the late blight in potato (*Phytophthora*) caused a famine in Ireland in 1845 (Fig. 6.8). Some examples of fungal vascular wilts also include *Fusarium* and *Verticillium* wilts in tomato.

Additionally, **phytopathogenic bacteria** can cause plant diseases and spread by infected seeds, irrigation, and other equipment. Some examples of bacterial disease also include *Erwinia, Xylella, and Xanthomonas*. In the case of bacteria, there must be a natural opening for bacteria to enter the plant tissue. Moreover, bacterial pathogens are also dependent on water such as wet environment conditions.

Table 6.2 Biofertilizer examples, their origin, and beneficial usage

Biofertilizers (natural, organic, eco-friendly)	Type and beneficial usage
1. Rhizobium biofertilizer	*Bacteria* (N-fixing for legumes)
2. Azotobacter biofertilizer	*Bacteria* (N-fixing, no plant symbiosis)
3. Mycorrhizae (VAM) biofertilizer	*Fungi* (enhances P, Zn uptake)
4. Cyanobacteria biofertilizer	*Algae* (N-fixing)
5. Azolla biofertilizer	*Fern* (N-fixing for rice)
6. Phosphobacteria biofertilizer	*Bacteria* (P solubilization)
7. Compost biofertilizer	Decomposed plant material

Viruses (over 1000) can also cause plant diseases with symptoms such as leaf rolls, ringspots, and deformed fruits. Some examples of viral disease also include mosaic patterns on leaves. Viruses must have a vector to be transmitted.

In the case of pests, insects can also cause plant diseases via piercing or sucking. Some examples of insects that cause diseases in plants also include vectors such as aphids, thrips mealybugs, caterpillars, grasshoppers, and beetles (Fig. 6.8).

Some of the management of plant diseases include:

- Defense: +Structural defenses (cuticle, cell wall), +biochemical defenses (toxic, or proteins such as salicylic acid (SA))
- Usage of pathogen-free seeds and plants
- Controlling insects
- **Crop rotation**
- Removing infected plant parts
- Usage of resistant plant varieties (or GMO cultivars)
- Biological control via usage of natural predators to fight diseases
- For more on this topic, see Martens et al. (1985), Ellis and Bradley (1992), and many references therein.

The diagnosis tips to keep your plants healthy are as follows:

- **Curling** on leaf: Aphids or thrips
- **Holes** on leaf: Caterpillars, ants, snails, slugs, or grasshoppers
- **Patches** on leaf: Downy mildew or powdery mildew
- Leaf scorch **browning on edge**: Too much sun, cold, or fertilizer
- **Spots** on leaf: Bacteria, fungi, or virus
- Leaf **yellowing**: Fe deficiency or N deficiency (not a disease)

For more on this topic, see Kochian et al. (2004), Taylor (2020), Martens et al. (1985), Marschner (1995), Balick and Cox (1996), Eichhorn et al. (2012), Reece et al. (2014), FAO (2017, 2024), FAOSTAT (2023), Hacisalihoglu (2011, 2021, 2022), Harshberger (1896), Judd et al. (2015), Levetin and McMahon (2020), Prance

Fig. 6.8 Four major groups of plants pathogen sources, diseases, their typical symptoms, representative plant species, and suggested solutions

et al. (2005), Ridolfi (2019), Schmidt and Cheng (2017), Simpson and Ogorzaly (2013), USDA (2023), USDA-NRCS (2000), Weber et al. (2024), Baysal et al. (2008), and many references therein.

6.4 Plant Profiles Bingo: How Many Can You Identify?

(P51)  *Ficus carica*	(P52)  *Fortunella japonica*
(P53)  *Fouquieria splendens*	(P54)  *Gerbera jamesonii*
(P55)  *Ginkgo biloba*	(P56)  *Gossypium herbaceum*
(P57)  *Guzmania sp.*	(P58)  *Gymnocalycium mihanovichii*
(P59)  *Helianthus annuus*	(P60)  *Hibiscus sabdariffa*

6.5 Takeaway Questions

1. What are the most crucial factors to consider when choosing plants?
2. What do plants use nitrogen for mainly?
3. Where will immobile nutrient deficiency symptoms show?
4. Which of the following is an effective method to control the spread of plant diseases?

6.6 Now You Know the Answers

1. Light, water, warm temperature, and nutrients
2. Synthesizing proteins and nucleic acids
3. In new growth part of the plant
4. Removing infected plant parts

References

Arnon DI, Hoagland DR (1940) Crops production in artificial solution and in soils with special reference to factors affecting yields and absorption of inorganic nutrients. Soil Sci 50:463–484

Balick MJ, Cox PA (1996) People, plants, and culture: the science of ethnobotany, New York

Baysal O, Calıskan M, Yesilova O (2008) An inhibitory effect of a new *Bacillus subtilis* strain (EU07) against *Fusarium oxysporum* f. sp. radicis-lycopersici. Physiol Mol Plant Pathol 73:25–32

Bewley JD, Bradford K, Hilhorst H, Nonogaki H (2013) Dormancy and the control of germination. In: Seeds, physiology of development, germination and dormancy, 3rd edn. Springer, New York, pp 247–297

Black M, Bewley JD (2000) Seed technology and its biological basis. CRC Press LLC, Boca Raton

Bradford KJ, Nonogaki H (2007) Seed development, dormancy and germination. Blackwell Publishing, Hoboken, pp 1–367

Dragland S et al (2003) Several culinary and medicinal herbs are important sources of dietary antioxidants. J Nutr 133:1286–1290

Eichhorn SE et al (2012) Raven biology of plants, 7th edn. Macmillan

Ellis BA, Bradley FM (1992) The organic gardener's handbook of natural insect and disease control. Rodale Press, Emmaus

FAO (2017) The future of food and agriculture: trends and challenges. Food and Agriculture Organization of the United Nations, Rome

FAO (2024) Food and Agriculture Organization of the United Nations. Current market situation and medium-term outlook 1 August 2025. http://www.fao.org

FAOSTAT (2023) FAO statistical databases. Food and Agriculture Organization of the United Nations: Rome, Italy; https://www.fao.org/faostat

Galuzzi G et al (2010) Home gardens: neglected hotspots of agro-biodiversity and cultural diversity. Biodivers Conserv 19:3635–3654

Hacisalihoglu G (2011) Plant biology concepts and laboratory manual. Kendall Hunt Publishing, Iowa

Hacisalihoglu, G. (2021). From growing to biology: plants 1st edn. Florida State University. 157

Hacisalihoglu G (2022) Level up college success: research, writing, and more key skills for STEM learners. ISBN13: 9798218066260. www.amazon.com/author/drhaci
Harshberger JW (1896) The purposes of ethnobotany. Bot Gaz 21:146–154
Ibanez A, Garrido-Chamorro S, Vasco-Cárdenas M, Barreiro C (2023) From lab to field: Biofertilizers in the 21st century. Horticulturae 9:1306
Judd WS et al (2015) Plant systematics: a phylogenetic approach, 4th edn. Sinauer Associates, Sunderland
Kochian LV et al (2004) How do crop plants tolerate acid soils? Mechanisms of aluminum tolerance and phosphorous efficiency. Annu Rev Plant Biol 55:459–493
Levetin E, McMahon K (2020) Plants and society, 8th edn. McGraw Hill
Marschner H (1995) Mineral nutrition of higher plants. Academic Press, London
Martens JW, Seaman WL, Atkinson TG (1985) Diseases of field crops in Canada. The Canadian Phytopathological Society, Harrow
Prance GT et al (2005) The cultural history of plants. Routledge, New York
Reece JB et al (2014) Campbell biology, 10th edn. Pearson, Boston
Ridolfi R (2019) SDGs for people, planet, and prosperity. http://www.openknowledge.fao.org/
Schmidt BM, Cheng DMK (2017) Ethnobotany: a phytochemical perspective. Wiley
Simpson BB, Ogorzaly MC (2013) Plants in our world: economic botany, 4th edn. McGraw-Hill
Taylor AG (2020) Seed storage, germination, quality, and enhancements. In: The physiology of vegetable crops, pp 1–30
USDA (2023) National Agriculture Statistics Service. https://www.nass.usda.gov/
USDA-NRCS (2000) The PLANTS database. National Resources Conservation Service. https://plants.usda.gov/
Weber C, Simnitt S, Wakefield H, Wechsler S, Swearingen B (2024) Fruit and tree nuts outlook: March 2024; United States Department of Agriculture: Washington, DC, USA. https://www.ers.usda.gov/publications/pub-details?pubid=108880

Chapter 7
Food Plants: What's for Dinner?

Food Plants: What's for Dinner?

Contents

G. Hacisalihoglu, *Super Greens*, https://doi.org/10.1007/978-3-032-17047-7_7

7.1 Big Picture: Quick Read Chapter Infographic

7.2 Student Learning Outcomes (SLOs) of the Chapter

- Evaluate economic impacts of food crop production in global contexts.
- Distinguish the differences among annual crop productions.
- Analyze the significance of corn, soybeans, wheat, and legumes.
- Define the role of food crops in global food security and nutrition.
- Understand the diversity and germplasm storage of crop plants (Fig. 7.1).

7.3 Concepts of the Chapter

7.3.1 Sustainable Development Goals (UN SDGs)

What are **SDGs** and what kind of positive difference could they make in communities and the world? In 2015, a set of **17 Sustainable Development Goals** (SDGs) was adopted by the United Nations (UN 2015) to achieve peace and prosperity for people and the planet as follows: No poverty, Zero hunger, Good health and well-being, Quality Education, Gender equality, Clean water and sanitation, Affordable and clean energy, Decent work and economic growth, Industry, innovation and infrastructure, Reduced disparity, Sustainable cities and economies, Responsible consumption and production, Environment action, Life below water, Life on land, Peace, Justice and strong institutions, and Partnership for the goals (Fig. 7.2).

The target deadline is set for the year 2030. Plant biology directly supports several SDGs, including enhancing food security and maximizing food production (SDG2), improving human health and well-being with green spaces (SDG3), sustainability and protection of green spaces (SDG11), environmental protection and health (SDG13), contribution to biodiversity (SDG15), and many others (Fig. 7.2).

It is the ultimate goal that SDGs-aligned research will more likely **drive real-world change** and impact policies worldwide. For more on this topic, see Ridolfi (2019) and many references therein.

For more on this topic, see UN (2015) and many references therein.

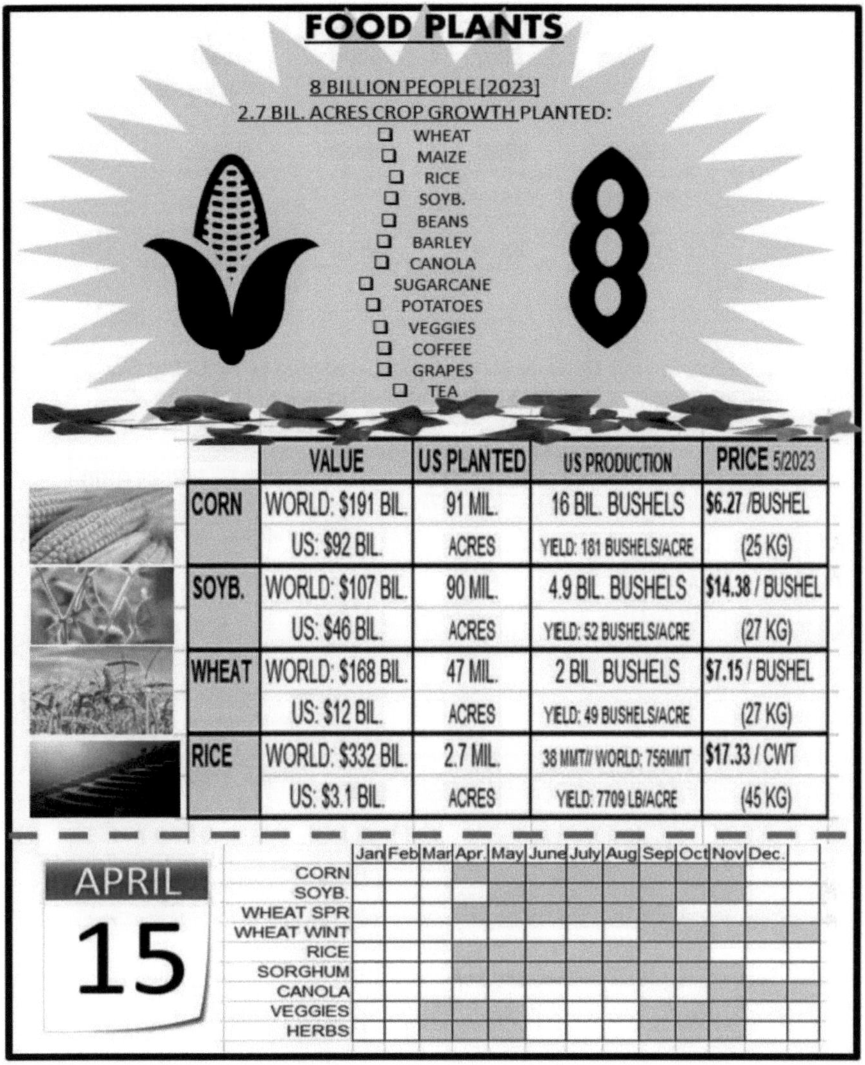

Fig. 7.1 Infographics of this chapter

7.3.2 *Global Food Security and Sustainability: Major Food Crop Species*

The human population is nearing about **10 billion by 2050**, meaning there will soon be more food needed (Fig. 7.3). Global food security is related to food access, availability, utilization, and sustainability for meeting populations' nutritional needs. As of year 2025, it is estimated that hundreds of millions of people are considered

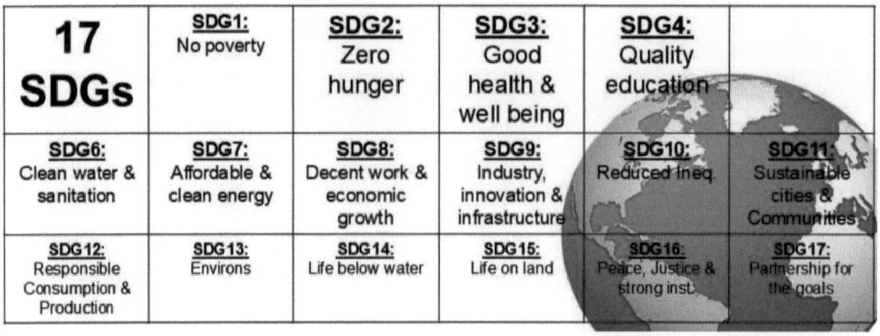

Fig. 7.2 The 17 Sustainable Development Goals (SDGs) adopted by the United Nations (UN) to achieve peace and prosperity for people and the planet (Data source: UN 2015)

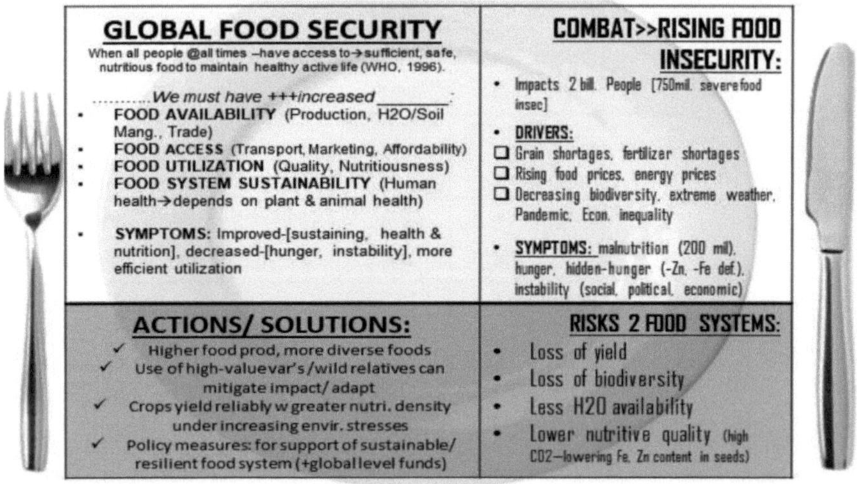

Fig. 7.3 Summary of global food security, combatting rising food insecurity, risks to food systems, and suggested solutions for the increasing world population

under malnutrition globally. Some of the major risk factors to food systems are as follows: Plant yield loss, biodiversity loss (lowers resilience), and lowered nutrition quality in plants (Fig. 7.3).

What are some suggested solutions? Plant breeding is utilized to develop new cultivars for extreme environmental stresses, cold or high temperatures, disease resistance, and low nutrient soil conditions. Moreover, adoption and promotion of more sustainable and higher diversity food crop varieties with superior nutrient levels could help improve global food security for next generations (Fig. 7.3).

For more on this topic, see McDonald (2010) and many references therein.

7.3.3 *Conservation of Crop Diversity and Germplasm Storage*

What are **genebanks** anyway? Seed (germplasm repositories) banks (aka genebanks) store and preserve crop seeds and other live plant parts for long-term availability and research purposes. Their main goal is to safeguard biodiversity and preserve seeds' viability for next generations. Some major features of genebanks include as follows:

- The world's largest backup genebank: Svalbard Global Seed Bank in Norway (aka Doomsday vault) for securing the duplicates of world food crop seeds as a backup. As of 2025, it stores 1.3 million seed samples (825,000 varieties of 6000+ crop species stored at −18 C securely).
- The world's second largest genebank is **US National Plant Germplasm System** (USNGPS) with 600,000 different accessions in 20 seed banks.
- Other large international genebanks include CIMMYT: Int. Maize & Wheat Improv. Center; IRRI: Int. Rice Res. Inst.; and CIAT: Int. Center for Tropical Agriculture (Fig. 7.4).
- The Svalbard genebank is considered a backup for all genebanks worldwide, so that the world's biodiversity will be safeguarded in the case of loss of seeds from other locations.

For more on this topic, see Fowler (2016) and many references therein.

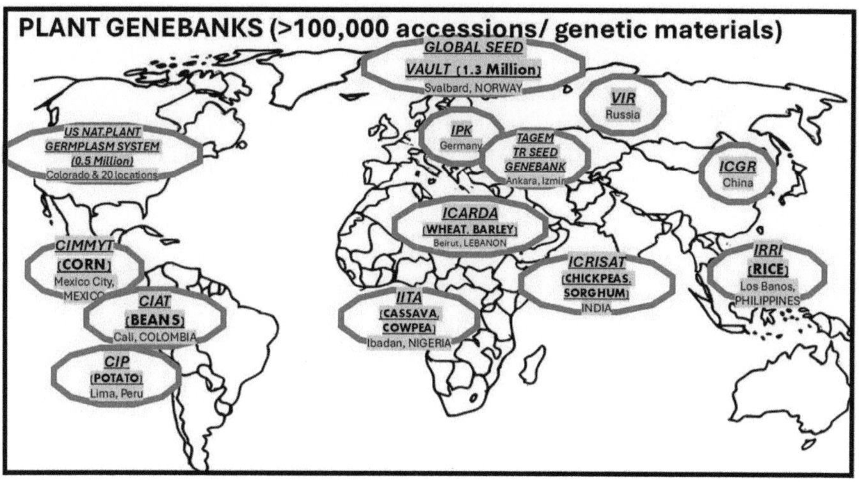

Fig. 7.4 Global largest plant germplasm repositories (genebanks) that store over 100,000 seeds, accessions, or genetic material. For more on this topic, see FAOSTAT (2023) and many references therein

7.3.4 Agricultural Yield

Agricultural **yield** is about the amount of crop produced per unit of land usage as well time (unit as tons per hectare). Yield is impacted by genetic factors, geolocation, weather, and crop species or varieties (Fig. 7.5).

For more on this topic, see Fischer et al. (2014), Smith and Hamel (2012), and many references therein.

7.3.5 Annual World Crop Plant Production

Food crop production is needed to be sustainably increased to feed the more populous world. Top 21 food crops, their market price (as of 2025), their US production, economic value, worlds production, area harvested, and major exporters countries are summarized in Table 7.1.

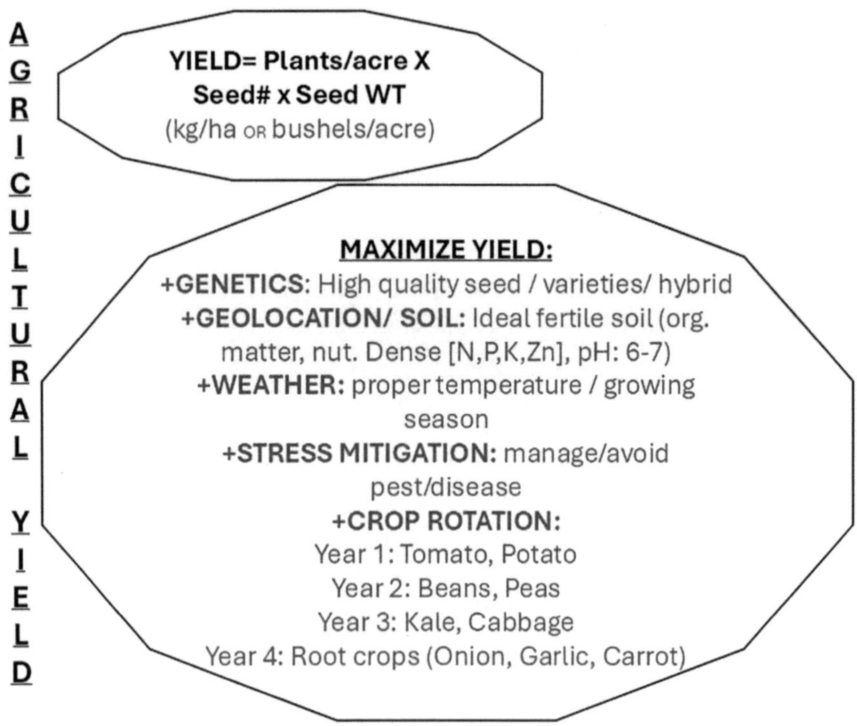

Fig. 7.5 Agricultural yield components and how to maximize yield

Table 7.1 Summary of US and global crop production, value, and market price by October 2025. Bushel: 25.4 kg (corn, sorghum), 27.2 kg (soy, wheat, peas), 22.7 kg (canola), 21.8 kg (barley), and 14.5 kg (oats, rye). For more on this topic, see FAOSTAT (2023) and many references therein

Crop	Market price ($) (Oct'25 wholesale)	US production	US value	World production (metric tons)	World value	World area (hectares)	Major exporters
Corn	4.00/bushel	15 Bil. bushels	**$90 Bil.**	1.1 Bil. MT	$250 Bil.	190 Mil. ha	USA, China, Brazil
Soybeans	10.17/bushel	4.6 Bil. bushels	**$50 Bil.**	380 Mil. MT	$150 Bil.	120 Mil. ha	USA, Brazil, Arg., China
Wheat	5.36/bushel	1.8 Bil. bushels	**$13 Bil.**	760 Mil. MT	$230 Bil.	220 Mil. ha	China, USA, Russia, EU
Tomato	0.5/lb	12 Mil MT	**$13 Bil.**	200 Mil. MT	$194 Bil.	5 Mil. Ha	China, Türkiye, USA, EU
Cotton	0.64/lb	15 Mil. bales	**$7 Bil.**	25 Mil. bales	$42 Bil.	31 Mil. ha	USA, India, China
Potatoes	0.89/lb	4.9 Bil. lb.	**$5 Bil.**	376 Mil. MT	$123 Bil.	17 Mil. ha	China, India, Ukraine, Russia
Rice	0.50/lb	12 Mil. MT	**$4.2 Bil.**	740 Mil. MT	$50 Bil.	160 Mil. ha	India, Vietnam, Thailand
Sorghum	4/bushel	8.7 Mil. MT	**$2.5 Bil.**	61 Mil. MT	$22 Bil.	41 Mil. ha	USA, Nigeria, Brazil
Canola	12.85/bushel	2.3 Mil. MT	**$2.5 Bil.**	75 Mil. MT	$37 Bil.	35 Mil. Ha	Canada, China, India, EU
Peas	10–16/bushel	0.3 Mil. MT	**$1.7 Bil.**	13 Mil. MT	$67 Bil.	25 Mil. Ha	USA, NL, Peru
Peanuts	0.26/lb	6.5 Bil. lb	**$1.6 Bil.**	47 Mil. MT	$93 Bil.	0.8 Mil. ha	China, India, Nigeria, USA
Barley	4/bushel	144 Mil. bushels	**$1.4 Bil.**	150 Mil. MT	$11 Bil.	50 Mil. Ha	EU, Australia, Ukraine
Beans	1.99/lb	9.9 Mil. acres	**$1 Bil.**	28 Mil. MT	$7 Bil.	33 Mil. Ha	USA, Canada, India, Brazil

(continued)

Table 7.1 (continued)

Crop	Market price ($) (Oct'25 wholesale)	US production	US value	World production (metric tons)	World value	World area (hectares)	Major exporters
Sunflower	0.22/lb	2.3 Bil. lb	**$419 Mil.**	50 Mil. MT	$41 Bil.	25 Mil. Ha	Ukraine, Russia
Oats	3.52/bushel	68 Mil. bushels	**$196 Mil.**	20 Mil. MT	$6.6 Bil.	5.6 Mil. Ha	Canada, AUS, Finland
Flaxseed	0.86/lb	2.7 Mil. bushels	**$142 Mil.**	875 K MT	$860 Mil.	2.3 Mil. ha	Canada, Russia
Chickpeas	0.43/lb	66 K MT	**$70 Mil.**	18 Mil. MT	$12 Bil.	14 Mil. ha	India, Australia, Canada
Rye	4.50/bushel	12 Mil. bushels	**$60 Mil.**	20 Mil. MT	$1.6 Bil.	10 Mil. ha	Poland, Germany, Canada
Coffee	2.93/lb	12 K MT	**$54 Mil.**	10 Mil. MT	$200 Bil.	10 Mil. ha	Brazil, Vietnam, Colombia
Safflower	0.64/kg	99 Mil. lb	**$52 Mil.**	724 Mil. MT	$138 Mil.	0.3 Mil. ha	Kazakhstan, Russia, NL
Tea	2.50/kg	–	–	6 Mil. MT	$40 Bil.	3 Mil. ha	China, India, Kenya, Sri Lanka, Türkiye

FOOD CROP/ VEGETABLE	TOP RANKING STATES & ECONOMIC CONTRIBUTION
Cabbage	1] California ($150 Mil.)
Corn	1] Iowa ($15 Bil.); 2] Illinois ($14 Bil.); 3] Nebraska ($11 Bil.)
Corn, sweet	1] Florida ($220 Mil.); 2] California ($178 Mil.); 3] Washington ($115 Mil.)
Cucumber	1] Florida ($150 Mil.); 2] Georgia ($80 Mil.)
Lettuce	1] Arizona; 2] California ($2 Bil.)
Onions	1] California ($400 Mil.); 2] Washington ($75 Mil.)
Peas	1] Wisconsin ($40 Mil.); 2] Minnesota ($30 Mil.)
Pepper, sweet	1] Florida ($100 Mil.)
Potatoes	1] Idaho ($1.2 Bil.); 2] Washington ($800 Mil.); 3] Maine ($160 Mil.)
Snap beans	1] Florida ($200 Mil.); 2] Wisconsin ($100 Mil.)
Soybeans	1] Illinois ($10 Bil.); 2] Iowa ($9 Bil.); 3] Minnesota ($5 Bil.)
Tomatoes	1] Florida ($500 Mil.); 2] California ($500 Mil.); 3] Ohio ($115 Mil.)
Wheat	1] N. Dakota ($2 Bil.); 2] Kansas ($2 Bil.); 3] Montana ($1 Bil.)

Fig. 7.6 Top ranking of the US states for major crop and vegetable species and their economic contribution

For more on this topic, see Strable and Scanlon (2009), Fowler (2016), Smith et al. (2014), Grusak and DellaPenna (1999), Hacisalihoglu (2024), Hacisalihoglu et al. (2021, 2022, 2025), Hacisalihoglu and Armstrong (2022), Taylor et al. (1998), Hacisalihoglu and Settles (2013, 2017), and many references therein.

A number of US states lead the agricultural production, including California, Iowa, Florida, and Texas (Fig. 7.6). For example, Florida is a leading producer of fresh market tomatoes, while Illinois leads pumpkin production.

7.3.6 *Poaceae Family (Grasses)*

Poaceae (grass family) includes over 10,000 species with **cereal crops** for food and sugar, forage crops, biofuels, and building materials sources (Fig. 7.7). They are in **monocotyledons class** with caryopsis type fruit (seed). Warm season Poaceae species carry out C4 photosynthesis pathway (corn, sorghum, sugarcane, millet), while cool season species carry out C3 photosynthesis pathway (wheat, rice, oats, rye).

For more on this topic, see Strable and Scanlon (2009), Fowler (2016), Smith et al. (2014), Grusak and DellaPenna (1999), Hacisalihoglu (2024), Hacisalihoglu et al. (2021, 2022, 2025), Hacisalihoglu and Armstrong (2022), Taylor et al. (1998), and many references therein.

Part 1: Corn (*Zea mays*): Corn (maize) is the most produced crop plant in the world (Figs. 7.8 and 7.9). It takes about 100 days from planting to harvest. Planting dates of corn starts on April 15 in the southern USA, while April 28 in northern USA. There are three major varieties of corn as follows:

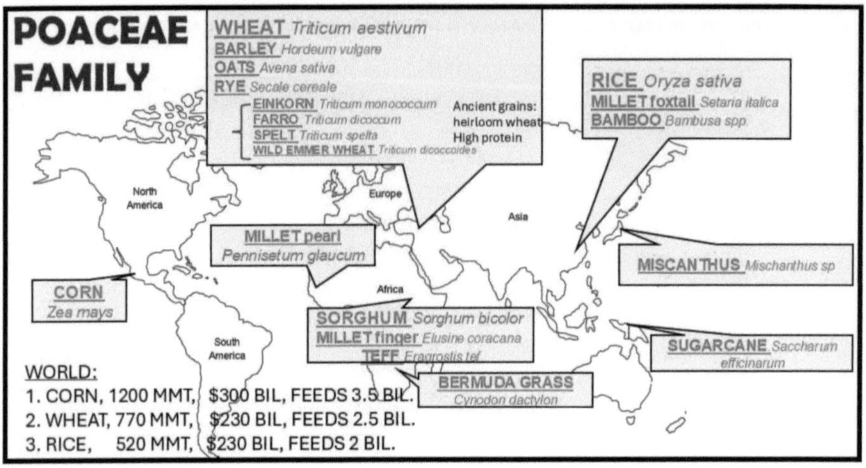

Fig. 7.7 Poaceae family (grass family) contains over thousands of species, including wheat ancestors einkorn and emmer. For more on this topic, see FAOSTAT (2023) and many references therein

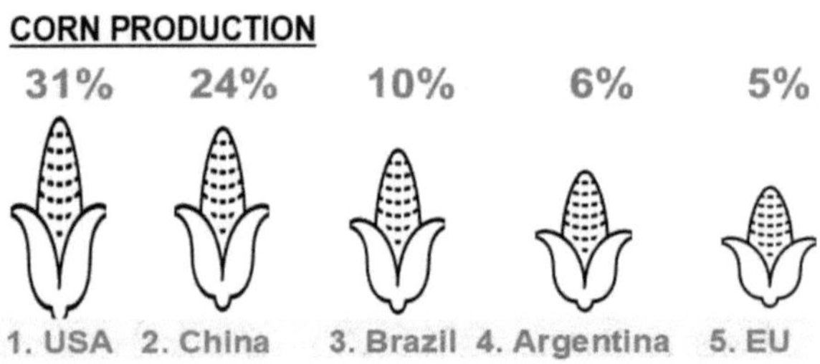

Fig. 7.8 Global corn production with largest leading producers (*Zea mays*)

Fig. 7.9 (**a**) Corn (*Zea mays*) plants; (**b**) corn ears; (**c**) corn kernels. (Image: G. Hacisalihoglu, PhD)

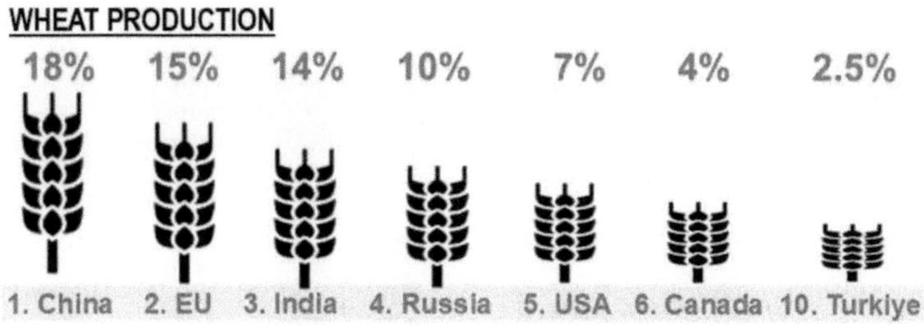

Fig. 7.10 Global wheat production with largest leading producers (*Triticum aestivum*). In the USA, Kansas, N. Dakota, and Montana are the leading wheat producing states

- **Dent corn** category (*Z. mays* var. indentata): Most modern corn: starch, feed
- **Popcorn** category (*Z. mays* var. everta): High water content kernels burst when heated
- **Sweet corn** category (*Z. mays* var. saccharata): High sugar content

Corn is the most productive and most energy efficient crop plant (5 tons per acre). Some industries of corn usage are as follows:

- Corn flour
- Corn starch
- Corn syrup / high-fructose
- Corn oil (4%)
- Corn alcohol (bourbon)
- Corn biofuel (ethanol)

For more on this topic, see Bennetzen and Hake (2009) and many references therein.

Part 2A: Wheat (*Triticum aestivum*): Wheat feeds over 3 billion people globally. The four major groups of wheat classifications are as follows (Figs. 7.10 and 7.11a):

- **Hard Red Spring Wheat** *(Triticum aestivum)*: High protein, gluten: Bagels, pizza
- **Hard Red Winter Wheat** *(Triticum aestivum)*: Bread
- **Soft Red Winter Wheat** *(Triticum aestivum)*: Starchy: Cookie, crackers, cakes
- **Durum Wheat** (*Triticum durum*): High gluten: Used for noodles

For more on this topic, see Anderson and Garlinge (2000) and many references therein.

Part 2B: Bread Making: Wheat was domesticated, and the first bread was made over 11,000 years ago in Fertile Crescent. Mills and commercial yeast accelerated bread making (Fig. 7.11b). Bread may be the oldest staple food globally. Leavened breads use yeast and produce CO2. Next, gluten protein takes CO_2 and raises the dough. There are several types of bread as follows:

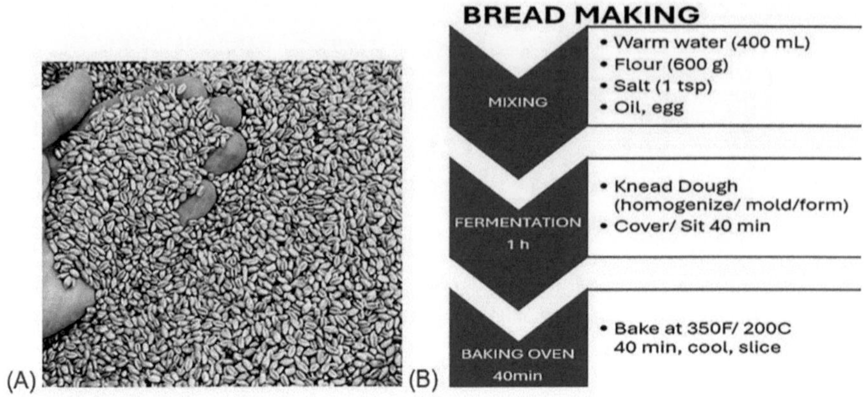

Fig. 7.11 (**a**) Wheat (*Triticum aestivum*) ranks third after corn and soybeans in the USA; (**b**) baking bread involves in mix, knead, rise, and oven bake of dough in 90 min time. (Image: G. Hacisalihoglu, PhD)

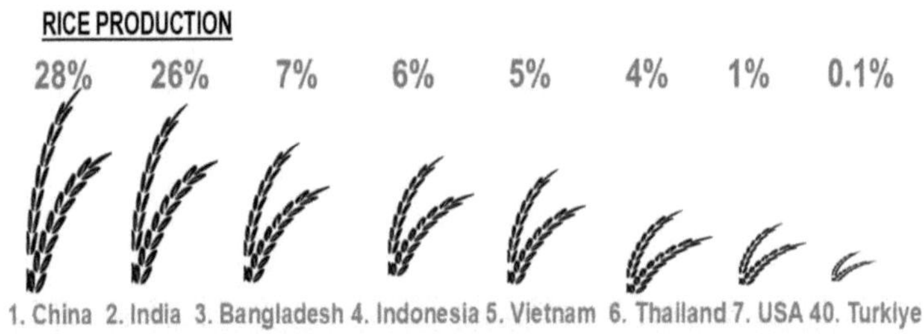

Fig. 7.12 Global rice production with largest leading producers (*Oryza sativa*)

- Baguette: Long, crisp, French bread loaf
- **100% Whole Grain**/Wheat: Spelt, flaxseed, chia seed, almond
- Sprouted Grain Bread: High fiber, nutrients
- Sourdough: Tangy, from wild yeast and bacteria (*Lactobacillus*), which produces sour lactic acid
- Pumpernickel: Rye bread
- Pita/Tortilla/Naan: Flatbread, some unleavened (no yeast)

Part 3: Rice (*Oryza sativa*): Rice feeds around 5 billion of people globally. (Figs. 7.12 and 7.13). Rice has two subspecies:

- **Indica rice** sub-sp (tropical): 6–8 mm long grain, fluffy: e.g., Basmati, Jasmine
- **Japonica rice** sub-sp (temperate): 4 mm short grain: Sushi rice, Sticky rice (sweet), nigiri, sashimi

While brown rice is whole grain with low GI, white rice (refined, polished, enriched) has high GI. Rice is a semiaquatic crop with two growth types:

Fig. 7.13 Rice (*Oryza sativa*) plants growing in Japan in semiaquatic rice fields (paddies, lowland rice). The US states of Arkansas, California, Louisiana, and Texas produce rice. (Image: G. Hacisalihoglu, PhD)

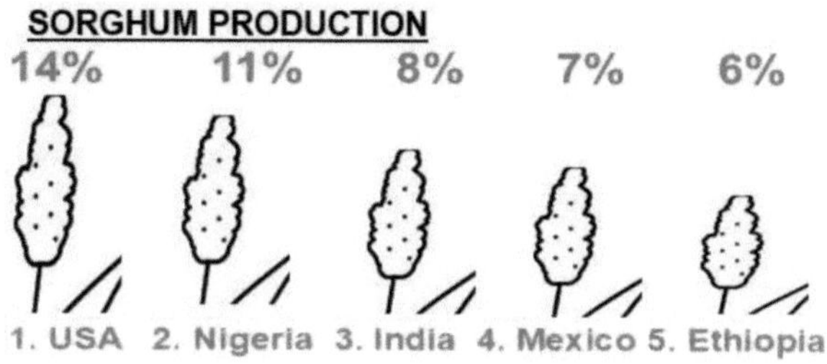

Fig. 7.14 Global sorghum production with largest leading producers (*Sorghum bicolor*)

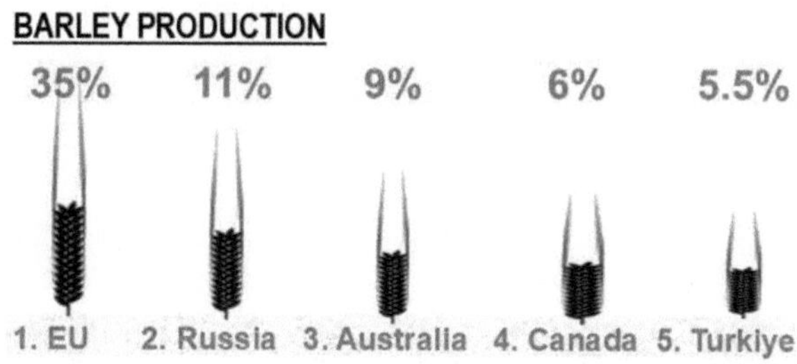

Fig. 7.15 Global barley production with largest leading producers (*Hordeum vulgare*)

- **Flooded** (lowland anaerobic cvs): 80% production: Paddies use 5–10 cm high water; high yield; Asia
- **Non-flooded** (upland drought tolerant cvs): Low yield; Tanzania, Guinea, Brazil

For more on this topic, see Bray et al. (2015) and many references therein.

Part 4: Sorghum (*Sorghum bicolor*): Sorghum is the fifth most important cereal crop, especially important in Africa and Asia (Fig. 7.14). As a climate-resilient crop, sorghum can help nourish world and improve global food security.

For more on this topic, see Zhao et al. (2019) and many references therein.

Part 5: Barley (*Hordeum vulgare*): Barley is one of the first domesticated plants and important for beverage, agriculture, and animal feed industries (Fig. 7.15). One of the key barley usage areas is malt production for beer.

For more on this topic, see Ullrich (2011) and many references therein.

Part 6: Oats (*Avena sativa*): Oat production and global importance are growing due to its health benefits (20 million metric tons, Fig. 7.16). Oats are known for their protein, minerals, vitamin, and fiber content. Some of the oat products include oatmeal, oat milk, oat flour, and oat bread. For more on this topic, see Welch (1995) and many references therein.

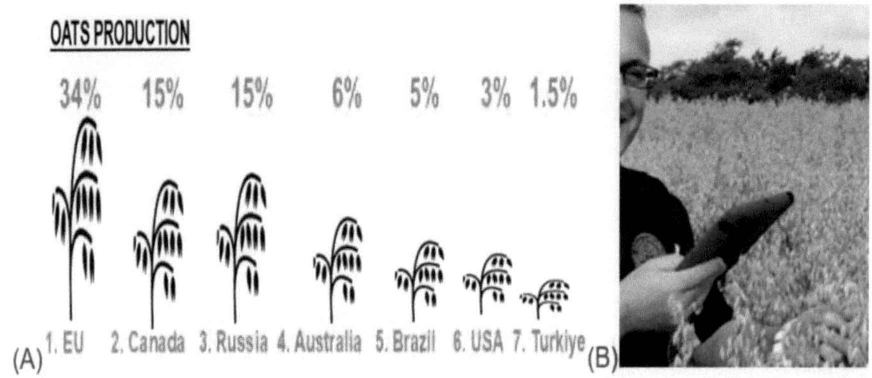

Fig. 7.16 (**a**) Global oat production with the largest leading producers (*Avena sativa*); (**b**) visiting oat fields near harvest time in Denmark. (Image: G. Hacisalihoglu, PhD)

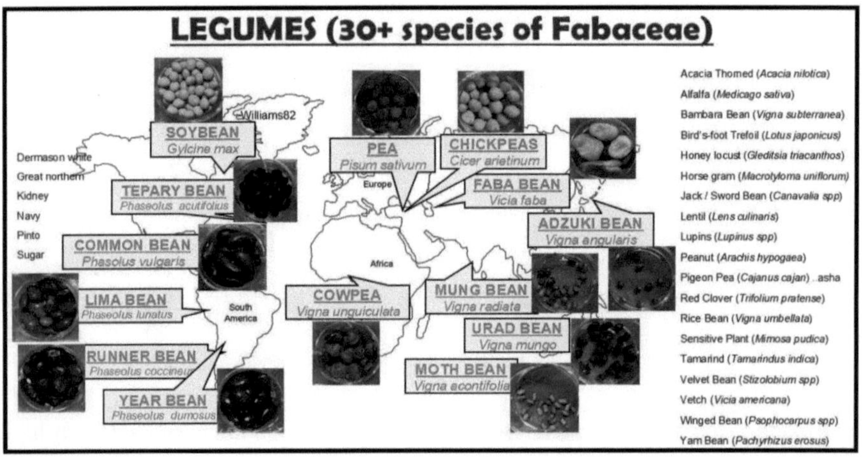

Fig. 7.17 Representative 30 major legume species summary of Fabaceae family out of 20,000 plus species. Pulses are the edible legumes for human diet, including common bean, soybean, chickpea, pea, lentil, and cowpea

7.3.7 Legumes (Soybean et al.)

Legumes feed over 300 million people globally. Legumes are rich sources of proteins due to their association with nitrogen fixing bacteria. Proteins play role in enzyme forming, hormones, and more. There are more than 20 K legume species globally (Fig. 7.17).

For more on this topic, see Kasai (2017), Hacisalihoglu (2024), Hacisalihoglu et al. (2021, 2022, 2025), Hacisalihoglu and Armstrong (2022), Taylor et al. (1998), Hacisalihoglu and Settles (2013, 2017), and many references therein.

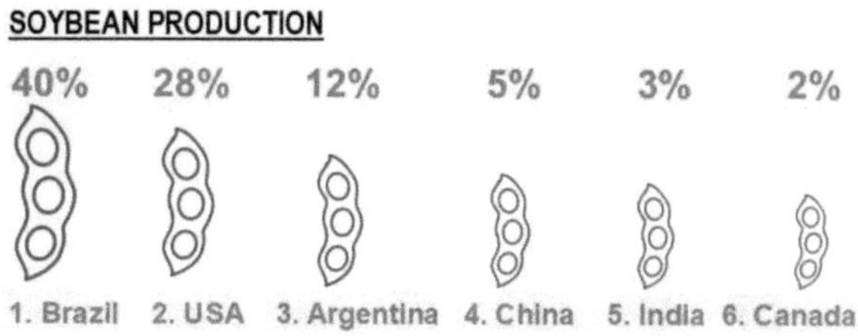

Fig. 7.18 Global soybean production with largest leading producers (*Glycine max*). In the USA (55% of soy production), Illinois and Iowa are the top soybean producers

Fig. 7.19 (**a**) Soybeans (*Glycine max*) are known for many uses such as tofu, miso, and soy oil; (**b**) common beans *(Phaseolus vulgaris)*; (**c**) peas (*Pisum sativum).* (Image: G. Hacisalihoglu, PhD)

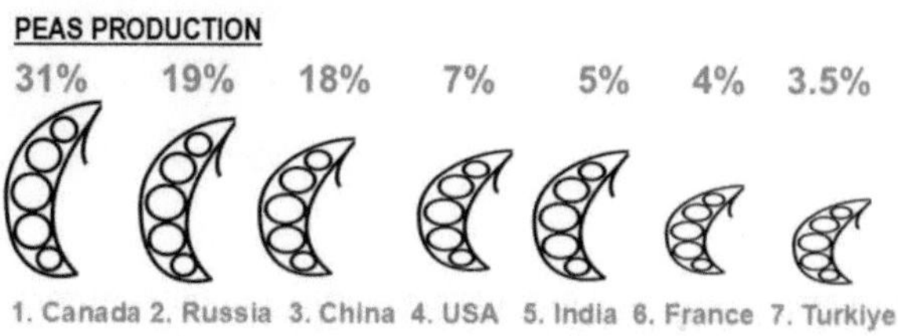

Fig. 7.20 Global pea production with largest leading producers (*Pisum sativum*)

Part 1: Soybeans (*Glycine max*): Soybeans are one of the most important legumes globally (Figs. 7.18 and 7.19). Soybeans are important for human diet, animal feed, and industrial usage, including **tofu (soy curd), miso (fermented soy), soymilk, soy sauce, edamame (fresh)**, soy flour, soy oil, soy sprouts (high vit. C content), soy proteins, soy drinks, plastics, tires, biofuel and more. Soybean tofu has 10 g protein per oz. For more on this topic, see Kasai (2017) and many references therein.

Part 2: Peas (*Pisum sativum*): Peas are cool season legumes and important plant-based protein globally (Figs. 7.19 and 7.20). Peas are rich in minerals such zinc and vitamins (vits. A and B).

For more on this topic, see Warkentin et al. (2015) and many references therein.

Part 3: Common Beans (*Phaseolus vulgaris*): Beans and grains (e.g., corn) are considered nutritionally complementary due to the fact that grains are high in methionine and cysteine, while legumes are low in them. (Fig. 7.21):

Legume domestication goes back to 7000 years for soybean (China), 6000 years for lentils, chickpeas, and **peas** (Türkiye), 4000 years for common beans (Peru). Pinto beans are named for painted beans, and navy beans are named after sailors.

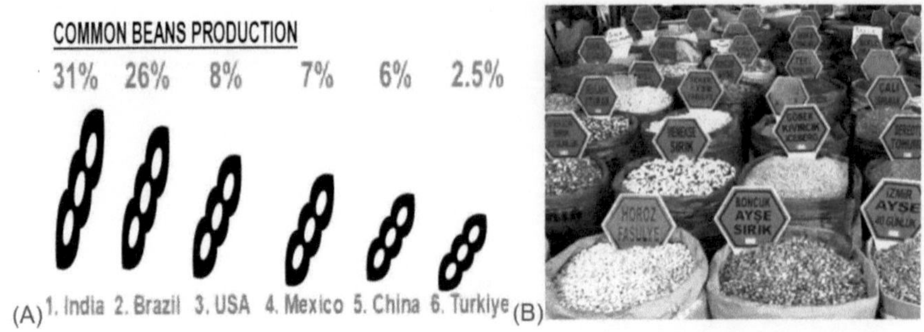

Fig. 7.21 (**a**) Global common bean production with largest leading producers. (*Phaseolus vulgaris*); (**b**) common bean varieties in a farmers' market. (Image: G. Hacisalihoglu, PhD)

Fig. 7.22 Chickpea (*Cicer arietinum*) varieties with light yellow color are preferred by consumers. One single chickpea plant can yield 100 seeds. (Image: G. Hacisalihoglu, PhD)

Common beans are highly rich in iron and potassium. Common beans are known for lowering cholesterol, blood pressure, as well as colon cancer. The starches in beans are healthy and known to increase good bacteria. Beans also help regulate blood sugar levels by preventing high spikes or crashes. For more on this topic, see Singh (1999) and many references therein.

Part 4: Chickpeas (*Cicer arietinum*): Chickpeas (aka garbanzo beans) are creamy textured legume species (Fig. 7.22). Chickpeas are rich in nutrients, proteins (8 g protein per oz), and dietary fibers. India and the USA (Montana, N. Dakota, Colorado, Oregon) are among the leading chickpea producers. For more on this topic, see Varshney et al. (2017) and many references therein.

7.3.8 Other Food Crops

Part 1: Potatoes (*Solanum tuberosum*): Potatoes feed around 5 billion people globally (Figs. 7.23 and 7.24). There are over **4000 potato varieties** in varied sizes, shapes, and colors. Potato arrived in Europe from South America in 1600. Although Ireland was most suitable for potato crops, ***Phytophthora* potato blight disease** caused the loss of 1.5 million people (Irish potato famine 1845).

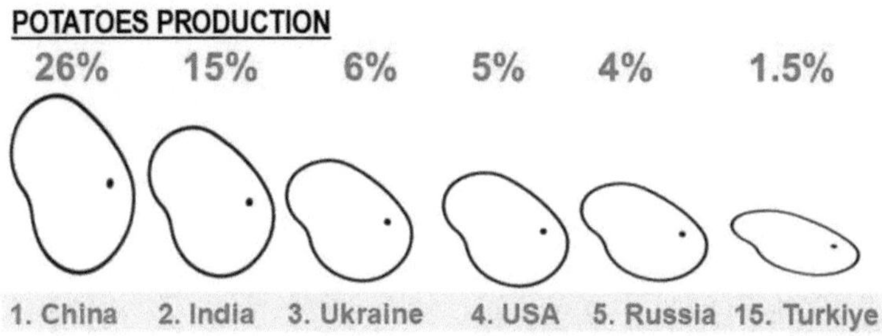

Fig. 7.23 Global potato production with largest leading producers. The USA grows 200 potato varieties in Idaho and Washington, including russet, red, white, yellow, and purple potatoes (Fig. 7.24)

Fig. 7.24 Potatoes (*Solanum tuberosum*) are starchy staple crops and domesticated in Andes/Peru 7000 years ago

For more on this topic, see De Jong et al. (2011) and many references therein.

Part 2: Sweet Potatoes (*Ipomoea batatas*): Sweet potatoes feed around millions of people globally (Fig. 7.25). Sweet potatoes are sweet tasting and thin skin tubers unlike yams (starchy tasting with brownish bark-like skin). The USA grows and exports sweet potatoes and North Carolina are the leading state for sweet potatoes. Sweet potatoes thrive in marginal soils and provide for nutritious vitamin A rich diet. For more on this topic, see Loebenstein and Thottappilly (2009) and many references therein.

Part 3: Tomato (S*olanum lycopersicum*): Tomatoes are very important staple foods worldwide (Figs. 7.26 and 7.27). Tomatoes are the **most cultivated and most consumed vegetable** globally. Tomatoes are consumed fresh and processed in soups and beverages.

For more on this topic, see Heuvelink (2018) and many references therein.

Fig. 7.25 Sweet potatoes (*Ipomoea batatas*). The leading global producers are China, Malawi, Tanzania, the USA (North Carolina). (Image: G. Hacisalihoglu, PhD)

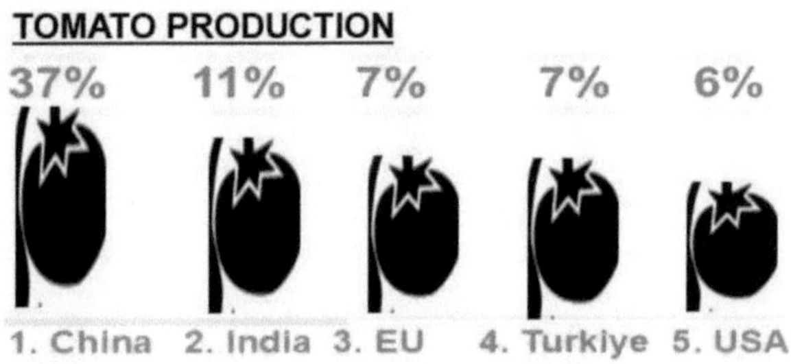

Fig. 7.26 Global tomato production with largest leading producers (*Solanum lycopersicum*)

Fig. 7.27 (**a**) Tomato (*Solanum lycopersicum*) plant; (**b**) tomato fruits. (Image: G. Hacisalihoglu, PhD)

7.3.9 *Oil Crops*

Oil crops supply essential fats, oils, proteins, and vitamins for diets. There are numerous oil crop species, including soybeans, palm, olive, canola, sunflower, peanut, cotton, and much more (Table 7.2).

Part 1: Canola (*Brassica napus*): Canola (aka **Canada's oil**) has been developed by plant scientists in the 1970s (Fig. 7.28). Today, canola is one of the most important oil seeds globally. For more on this topic, see Daun et al. (2015) and many references therein.

Part 2: Olives (*Olea europaea*): Olive trees are ideal choice for growers due to their low water usage, low pesticide requirement, simple harvest, and drought and heat tolerance (Fig. 7.29a).

For more on this topic, see Therios (2009) and many references therein.

Table 7.2 Sixteen oil crops, their heat smoke levels, potential health benefits, world production, and top producing countries. MUFAs, monounsaturated fats

Oil crops	Heat level	Health benefits	World production	Top producing countries
Avocado oil	High heat	Healthy-monounsaturated fat 70%	~	Mexico, DR, Peru
Olive oil (extra virgin)	No heat	Healthy-monounsaturated fat 77%	2 0.7 Mil. MT	Spain, Italy, Türkiye
Walnut oil	No heat	Healthy-rich in alpha-linolenic acid	~	USA, China
Pumpkin seed oil	No heat	Healthy-rich in zinc	~	Austria, Slovenia
Flaxseed oil	No heat	Healthy-rich in alpha-linolenic acid	1 Mil. MT	China, USA
Sesame oil	No heat	Healthy-but high calorie	1 Mil. MT	India, China
Canola oil	High heat	Monounsaturated 62%	87 Mil. MT	Canada, China
Palm oil	High heat	~	65 Mil. MT	Indonesia, Mal.
Soybean oil	High heat	~	53 Mil. MT	USA, Brazil
Sunflower oil	High heat	Polyunsaturated 69%	15 Mil. MT	Ukraine, Russia
Peanut oil	High heat	~	5 Mil. MT	China, India
Safflower oil	High heat	~	0.7 Mil. MT	USA, India
Coconut oil	High heat	Saturated 92%	3 Mil. MT	Philippines, Indo.
Grapeseed oil	High heat	Polyunsaturated 71%	0.5 Mil. MT	China, Italy
Cottonseed oil	No heat	~	4 Mil. MT	USA, India
Corn oil	No heat	~	4 Mil. MT	USA, China

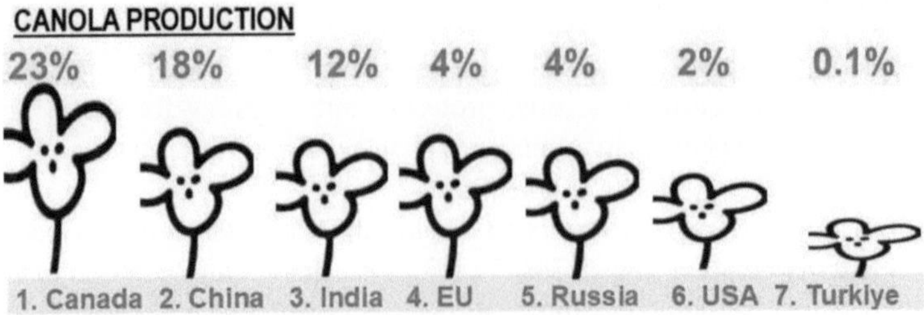

Fig. 7.28 Global canola production with largest leading producers (*Brassica napus*)

Fig. 7.29 (**a**) Olive tree (*Olea europaea*); (**b**) olive oils from Europe. Spain, Italy, and Türkiye are top olive oil producing countries. Olive oil is rich in monounsaturated fats (MUFAs), which are considered healthy fats in superfoods. (Image: G. Hacisalihoglu, PhD)

Part 3: Avocado (*Persea americana*): Avocado oil is known for its rich health benefits globally (Fig. 7.30). Avocado oil also has advantages of high smoke points to make it ideal food frying.

For more on this topic, see Miller (2020) and many references therein.

Part 4: Chocolate Production (*Theobroma cacao*): Did you know that chocolate grows on trees? Chocolate is produced from white beans of **cacao fruit** (football shaped pods) seeds (cacao beans). After fermentation processing of the cacao beans, they turn into complex brown and sweet chocolate flavor (Fig. 7.31). For more on this topic, see McNeil (2006) and many references therein.

Part 5: Flaxseed (*Linum usitatissimum*): Flaxseed is known for its rich nutrient content globally (Fig. 7.32). Flaxseed has high amounts of **ALA (omega-3 fatty acid alpha-linolenic acid)** content. Flaxseed is rich in protein (5 g protein per oz) and antioxidants.

For more on this topic, see Armstrong (2006), Charmet (2011), Egli (2017), Diamond (2002), Farah (2019), GRIN-Global (2017), Balick and Cox (1996), Eichhorn et al. (2012), Reece et al. (2014), FAO (2017, 2024), FAOSTAT (2023), Hacisalihoglu (2011, 2021, 2022), Harshberger (1896), Judd et al. (2015), Levetin

Fig. 7.30 Avocado and grape oils. (Image: G. Hacisalihoglu, PhD)

Fig. 7.31 The evergreen cacao tree (*Theobroma cacao*), its yellow fruit, and white edible cacao bean seeds are native to Central and South America. (Image: G. Hacisalihoglu, PhD)

Fig. 7.32 Flax (*Linum usitatissimum*) accessions with high oil and low oil content. (Image: G. Hacisalihoglu, PhD)

and McMahon (2020), Prance et al. (2005), Ridolfi (2019), Schmidt and Cheng (2017), Simpson and Ogorzaly (2013), USDA (2023), USDA-NRCS (2000), Weber et al. (2024), and many references therein.

7.4 Plant Profiles Bingo: How Many Can You Identify?

(P61) 
Arachis hypogaea

(P62) 

Phaseolus vulgaris

(P63) 
Fragaria ananassa

(P64) 
Pisum sativum

(P65) 
Glycine max

(P66) 
Triticum aestivum

(P67) 
Lactuca sativa

(P68) 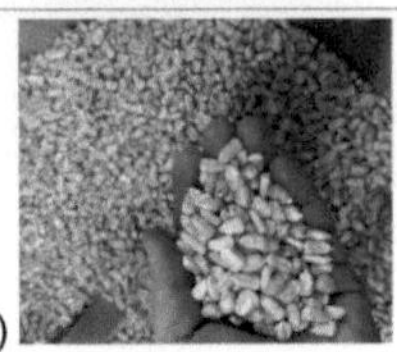
Zea mays

(P69) 
Sorghum bicolor

(P70) 
Solanum lycopersicum

7.5 Takeaway Questions

1. Define global food security.
2. What is the number one most grown cereal crop species worldwide.
3. What is the number one most oil crop species worldwide.
4. Which oil crop species is used to make canola?

7.6 Now You Know the Answers

1. All people have access to sufficient, nutritious food for their diets and healthy life.
2. Corn (maize).
3. Oil palm.
4. *Brassica napus.*

References

Anderson WK, Garlinge JR (2000) The wheat book: principles and practice. Department of Agriculture, Western Australia and Grains Research and Development Corporation

Armstrong P (2006) Rapid single-kernel NIR measurement of grain and oil-seed attributes. Appl Eng Agric 22:767–772

Balick MJ, Cox PA (1996) People, plants, and culture: the science of ethnobotany, New York

Bennetzen JL, Hake SC (2009) Handbook of maize: its biology. Springer Science & Business Media, New York

Bray F et al (2015) Rice: global networks and new histories. Cambridge University Press, New York

Charmet G (2011) Wheat domestication: lessons for the future. C R Biol 334:212–220

Daun JK, Eskin MN, Hickling D (2015) Canola: chemistry, production, processing, and utilization. Elsevier, Amsterdam

De Jong H et al (2011) The complete book of potatoes: what every grower and gardener needs to know. Timber Press, Portland

Diamond J (2002) Evolution, consequences and future of plant and animal domestication. Nature 418:700–707. https://doi.org/10.1038/nature01019

Egli DB (2017) Seed biology and yield of grain crops, 2nd edn. CABI International, Boston

Eichhorn SE et al (2012) Raven biology of plants, 7th edn. Macmillan

FAO (2017) The future of food and agriculture: trends and challenges. Food and Agriculture Organization of the United Nations, Rome

FAO (2024) Food and Agriculture Organization of the United Nations. Current market situation and medium-term outlook 1 August 2025. http://www.fao.org

FAOSTAT (2023) FAO statistical databases. Food and Agriculture Organization of the United Nations: Rome, Italy; https://www.fao.org/faostat

Farah A (2019) Coffee: production, quality and chemistry, 1st edn. The Royal Society of Chemistry, London

Fischer RA, Byerlee D, Edmeades GO (2014) Crop yields and global food security: will yield increase continue to feed the world? ACIAR Monograph No. 158; Australian Centre for International Agricultural Research: Canberra, ACT, Australia, 2014; Volume XXII, 634p

Fowler C (2016) Seeds on ice: Svalbard and the global seed vault. Prospecta Press, Westport

GRIN-Global (2017) U.S. National Plant Germplasm System. GRIN-Global. https://npgsweb.ars-grin.gov/gringlobal/search.aspx?

Grusak MA, DellaPenna D (1999) Improving the nutrient composition of plants to enhance human nutrition and health. Ann Rev Plant Physiol Plant MB 50:133–161

Hacisalihoglu G (2011) Plant biology concepts and laboratory manual. Kendall Hunt Publishing, Iowa

Hacisalihoglu G (2021) From growing to biology: plants, 1st edn. Florida State University, p 157

Hacisalihoglu G (2022) Level up college success: research, writing, and more key skills for STEM learners. ISBN13: 9798218066260. www.amazon.com/author/drhaci

Hacisalihoglu G (2024) Analysis of nutritional traits: natural variation within 90 diverse faba bean (*Vicia faba*) genotypes and daily value contribution. Crops 4(3):440–446. https://doi.org/10.3390/crops4030031

Hacisalihoglu G, Armstrong P (2022) Flax and sorghum: multi-element contents and nut values within 210 varieties & pot selection for future climates to sustain food security. Plants 11(3): 451. https://doi.org/10.3390/plants11030451

Hacisalihoglu G, Settles AM (2013) Natural variation of seed composition of 91 common bean genotypes and their possible assoc w seed coat color. J Plant Nutr 36(5):772–780. https://doi.org/10.1080/01904167.2012.754041

Hacisalihoglu G, Settles AM (2017) Quantification of seed ionome variation in 90 diverse soybean (*Glycine max*) lines. J Plant Nutr 40(20):2808–2817. https://doi.org/10.1080/01904167.2017.1382528

Hacisalihoglu G, Beisel N, Settles AM (2021) Characterization of pea seed nutrient value within a diverse population of *P.* sativum. PLoS One 16(11):e0259565. https://doi.org/10.1371/journal.pone.0259565

Hacisalihoglu G, Armstrong P, Mendoza T, Seabourn B (2022) Compositional analysis in sorghum (*Sorghum bicolor*) NIR spectral techniques based on mean spectra from single kernels. Front Plant Sci 13:995328. https://doi.org/10.3389/fpls.2022.995328

Hacisalihoglu G, White G, Salters V (2025) Characterization of seed quality Traits within 61 chickpea (*Cicer arietinum*) genotypes: ionomes and daily values. Seeds 4(1):15. https://doi.org/10.3390/seeds4010015

Harshberger JW (1896) The purposes of ethnobotany. Bot Gaz 21:146–154

Heuvelink EP (2018) Tomatoes, 2nd edn. CABI, Cambridge

Judd WS et al (2015) Plant systematics: a phylogenetic approach, 4th edn. Sinauer Associates, Sunderland

Kasai M (2017) Soybean-the basis of yield, biomass, and productivity. InTechOpen

Levetin E, McMahon K (2020) Plants and society, 8th edn. McGraw Hill

Loebenstein G, Thottappilly G (2009) The Sweet potato. Springer, Dordrecht

McDonald BL (2010) Food security. Polity Press, Cambridge

McNeil C (2006) Chocolate in Mesoamerica: a cultural history of Cacao. University Press of Florida, Gainesville

Miller J (2020) Avocado: a global history. Reaktion Books, London

Prance GT et al (2005) The cultural history of plants. Routledge, New York

Reece JB et al (2014) In: 10th edn (ed) Campbell biology. Pearson, Boston

Ridolfi R (2019) SDGs for people, planet, and prosperity. http://www.openknowledge.fao.org/

Schmidt BM, Cheng DMK (2017) Ethnobotany: a phytochemical perspective. Wiley

Simpson BB, Ogorzaly MC (2013) Plants in our world: economic botany, 4th edn. McGraw-Hill

Singh SP (1999) Common bean improvement in the twenty-first century. Kluwer Academic Publishers, Dordrecht

Smith DL, Hamel C (2012) Crop yield: physiology and processes. Springer, Berlin/Heidelberg, p 118

Smith JSC et al (2014) Yield gains in major U.S. field crops. Spec. Publ. 33. CSSA, Madison, WI

Strable J, Scanlon MJ (2009) Maize (*Zea mays*): a model organism for basic and applied research in plant biology. Cold Spring Harb Protoc 10:pdb. emo132

Taylor AG et al (1998) Seed enhancements. Seed Sci Res 8:245–256
Therios IN (2009) Olives. CABI, Oxfordshire
Ullrich SE (2011) Barley: production, improvement, and uses. World Agriculture Series. Wiley, Hoboken
UN (2015) *Transforming our world*: the 2030 Agenda for sustainable development. United Nations, New York
USDA (2023) National Agriculture Statistics Service. https://www.nass.usda.gov/
USDA-NRCS (2000) The PLANTS database. National Resources Conservation Service. https://plants.usda.gov/
Varshney RK, Thudi M, Muehlbauer FJ (2017) The chickpea genome. Springer, Cham
Warkentin TD et al (2015) Pea (*Pisum. sativum* L.). In: De Ron AM (ed) Grain legumes. Springer, New York, pp 37–83
Weber C, Simnitt S, Wakefield H, Wechsler S, Swearingen B (2024) Fruit and tree nuts outlook: March 2024; United States Department of Agriculture: Washington, DC, USA. https://www.ers.usda.gov/publications/pub-details?pubid=108880
Welch RW (1995) The oat crop: production and utilization. Chapman & Hall, London
Zhao ZY, Che P, Glassman K, Albertsen M (2019) Sorghum. Humana Press, New York

Chapter 8
Beverage Plants

Beverage Plants

Contents

G. Hacisalihoglu, *Super Greens*, https://doi.org/10.1007/978-3-032-17047-7_8

8.1 Big Picture: Quick Read Chapter Infographics

8.2 Student Learning Outcomes (SLOs) of the Chapter

- Understand plants used for beverages.
- Define tea cultivation and distinguish between various tea types.

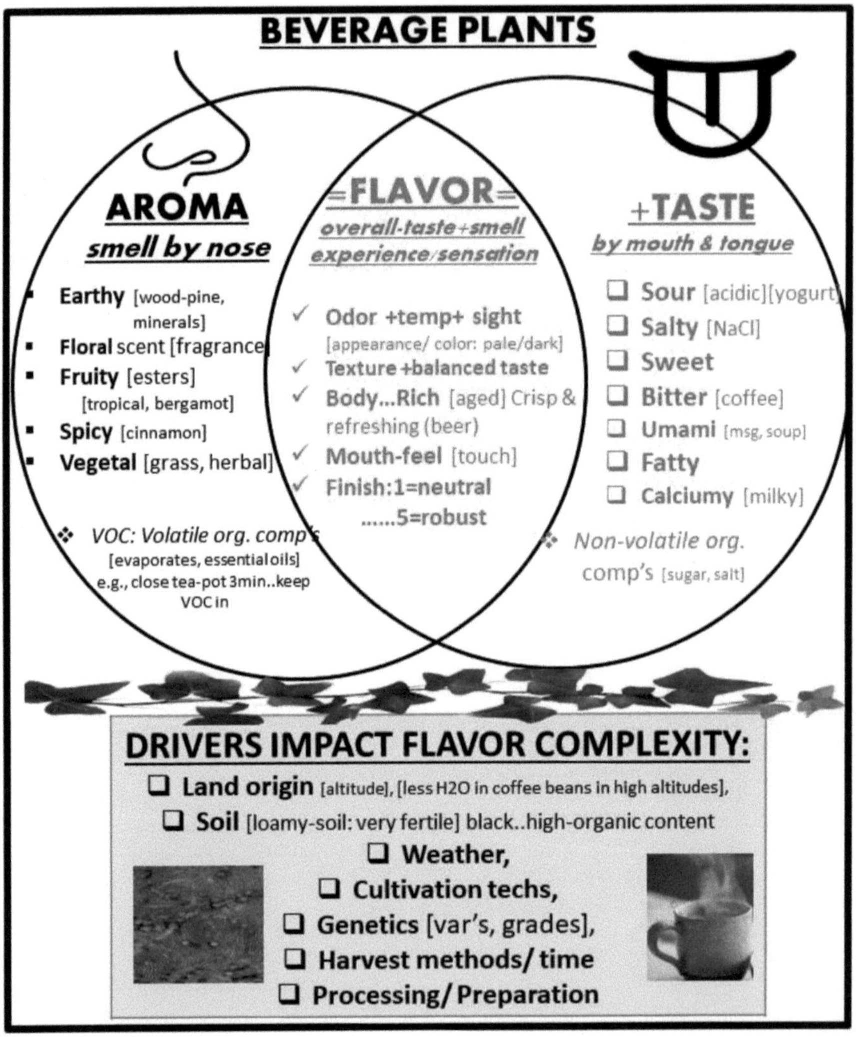

Fig. 8.1 Infographics of this chapter

- Define coffee cultivation and distinguish between various coffee types.
- Distinguish between flavor and taste and define them.

8.3 Concepts of the Chapter

8.3.1 Botany of Drinks

Except water and dairy beverages, plants are a major source of beverages. Plant beverages come from several plant species as follows:

- Tea plants (e.g., tea)
- Coffee plants (e.g., coffee)
- Cacao plants (e.g., chocolate)
- Sugarcane plants (e.g., rum)
- Juniper berries (e.g., gin)
- Various fruits (e.g., coconut)
- Various crop plants (e.g., corn [bourbon], oats, rice [sake], barley [beer, scotch], wheat, potatoes [vodka], grapes [wines], maize, agave [tequila], apples [ciders])
- Vegetables (e.g., tomatoes)

8.3.2 Coffee Beans: From Seed to Cup

Coffee (*Coffea arabica*), a cash crop with a $200 billion industry, is grown predominantly in coffee belt regions with warm temperatures and rain (Brazil to Ethiopia and Indonesia) globally. Today, coffee is one of the most widely traded commodities with $100 billion annual economic value. Coffee is mostly produced in small farms in 50 countries from harvest to drying, roasting, grinding, and finally brewing with various flavors and aromas. First coffee houses emerged in Istanbul around 1540s, and Türkiye introduced coffee to Europe later. Consequently, Europe introduced the coffee in Caribbean and South America. Coffee is made from seeds, which are found inside the fruit or coffee cherry (Figs. 8.2 and 8.3).

Some of the most valued coffee varieties are grown in the following locations:

- JAMAICA: Blue Mountain (less acidic)
- PANAMA: Geisha (fruity flavors), COSTA RICA (rich, volcanic soils)
- COLOMBIA: Ospina (luxury, handpicked beans, $3000/kg)
- GUATEMALA: Fina El Injerto (rich, full-bodied sweetness)
- KENYA (bold, earth flavors)
- ETHIOPIA (center of origin of coffee): Yirgacheffe coffee
- HAWAII (Big Island, volcanic soils): Kona coffee
- INDONESIA (Sumatra): Mandheling

Fig. 8.2 Global coffee production with largest leading producers. Caffeine is the alkaloid stimulant in coffee beans

Fig. 8.3 Coffee plants. Specifically, Arabica coffee varieties thrive in acidic soils (pH 4.5–6 levels), while Robusta varieties prefer slightly acidic soils (pH 5.5–6 levels). Higher altitudes and rainfalls of Costa Rica contribute to coffee flavor of fruit and spice. (Image: G. Hacisalihoglu, PhD)

- TURKISH COFFEE: Mehmet Efendi-aromatic fine ground powder from blended beans from South America.

For more on this topic, see Morris (2018), Illy (2002), Farah (2019), Bitter et al. (2020), Carlsen et al. (2010), and many references therein.

8.3.3 *Tea Leaves (Tea If by Sea, Chai If by Land)*

Only water is consumed more than tea globally. **Tea** (*Camelia sinensis*), a cash crop with a $40 billion market, is mainly grown in subtropical climates with plenty rainfall and acidic soil types in 66 countries. Global tea production is led by

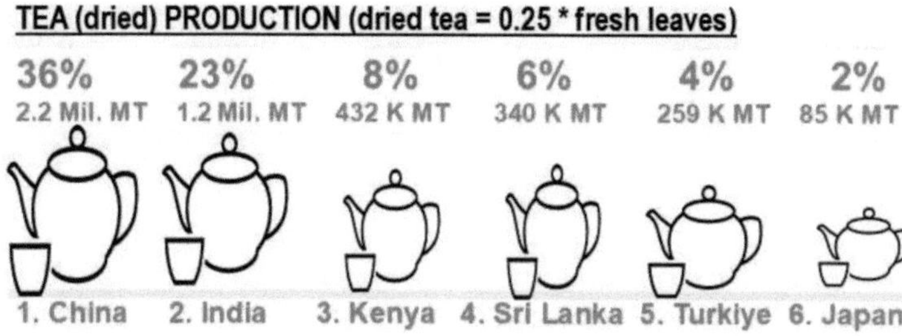

Fig. 8.4 Global tea production with largest leading producers

Fig. 8.5 Tea (*Camelia sinensis*) field and harvested tea leaf parts. Tea can be vegetatively propagated via single-leaf cuttings in 5 months. Tea can be harvested from 3 to 5 years mature plants and have 100 years lifespan. (Image: G. Hacisalihoglu, PhD)

China, India, Kenya, Sri Lanka, and Türkiye (Figs. 8.4, 8.5 and 8.6). High-quality tea leaves are commonly harvested in the following soil and weather conditions:

- High humidity
- Acidic, sandy, high organic matter soils
- Weather of the specific year (temperatures 16-30C), shade
- Region (high altitude→ more leaf aromatic compound and oils) and varieties/ cultivars

Tea has amazing health benefits: has high antioxidants, lowers cholesterol, lowers blood sugar, lowers blood pressure, is anti-inflammatory, and helps in weight loss, among others.

Tea leaf is the prime part of tea and harvested from top three leaves and bud (Fig. 8.5). High-end teas are commonly handpicked for selecting the best leaves. Polyphenols are the main metabolite of tea.

TEA CHEMISTRY / PROCESSING	❑ BLACK TEA -Caffeine: 45mg/ / 8 oz cup	❑ OOLONG TEA -Caffeine: 30mg/ / 8 oz cup	❑ GREEN TEA -Caffeine: 20mg/ / 8 oz cup	❑ WHITE TEA -Caffeine: 15mg/ 8 oz cup
1) Harvest /plucking	-pick 2-4 leaf +bud	-pick 2 leaf + bud	-pick 2 leaf + bud	-pick bud
2) withering	yes	yes	yes	yes
3) rolled	yes	yes	yes	No, just natural shape
4) allowed to oxidize	2-8 h oxidized to turn black color	partially	none	none
5) drying	medium	medium	low	low
***Camelia sinensis* varieties & cultivars**	var. *assamica* [large-leaf] var. *sinensis* *[in Turkiye→small leaf, cold-hardy]*	*C. sinensis* var. *sinensis*	*C. sinensis* var. *sinensis* China-type used only	Artisan tea
Best teas	English Breakfast Irish Breakfast Filiz, Rize cay (Turkiye)	Iron Goddess	Gunpowder (Chinese) Sencha (Japanese)	Silver needle Fujian, Taiwan
Taste	Robust, bold, malty	Floral, fruity	Vegetal, grassy	Light, floral

Fig. 8.6 Tea chemistry and processing with major tea types worldwide. Withering is the wilting of tea leaves to remove water. Rolling gives shape to tea leaves and enhances taste. Drying under heat stops oxidation and seals flavors

The main types of teas are white, green, oolong, and black. All true teas come from *Camelia sinensis*. Other factors that may affect tea flavor are various tea processing methods (Fig. 8.6). Full oxidization allows black teas to have their color and robustness.

For more on this topic, see McKay and Blumberg (2002), Richelle et al. (2001), and many references therein.

Part 1: Turkish Tea by Black Sea Türkiye has a rich history of tea culture. Turkish tea is mostly black tea. Black Sea region (of Trabzon and Rize) is particularly known for their high-quality teas. Turkish tea production mostly caters to domestic needs. Tea is the most common welcome drink. With over 4 kg/year, Türkiye has **the highest tea consumption** worldwide. The other countries with high tea consumption are China, India, and Argentina. For more on this topic, see Kacar (1992), Yazici (2024), and many references therein.

8.3.4 Grapes and Wine

Grapes (*Vitis* spp.) are divided into table or wine grape categories. Grapes are also divided into two groups: **white (W) and red (R) grapes** (Figs. 8.7 and 8.8). Seedless trait is important for table grapes, while high sugar content is important for wine grapes. First domestication of grapes goes back to 6000 years ago in Black Sea

Fig. 8.7 (**a**) Grape (*Vitis vinifera*); (**b**) red and white grapes; (**c**) white and red wine. (Image: G. Hacisalihoglu, PhD)

European-type (*Vitis vinifera*)	**POPULAR GRAPE VARIETIES (10,000+ var's)**
TABLE GRAPES fresh consumption, higher yield	▪ **Sultana** (Thompson seedless) (W) ▪ Cotton Candy (W) ▪ Muscat (W) ▪ Concord (R): most grown in Midwest ▪ Moon Drops (R)
WHITE (W) WINE GRAPES Smaller, thicker skin, seed large	▪ Riesling (W) ▪ **Chardonnay** (queen of (W) wines): most planted wine var. in the USA ▪ Airen (W) ▪ Sauvignon Blanc (W) ▪ Narince (W)
RED (R) WINE GRAPES Smaller, thicker skin, seed large	▪ **Cabernet Sauvignon** (cabs: king of (R) wines): long very late growing season: 2nd most planted ▪ **Pinot Noir** (R): third most planted ▪ Merlot (R) ▪ Tempranillo (R) ▪ Shiraz (R) ▪ Zinfandel (R) ▪ Sangiovese (R) ▪ Cabernet Franc (R): most planted in East

Fig. 8.8 Grape (*Vitis vinifera*) types. R: red; W: white. Grapes are harvested, crushed, fermented, filtered, and bottled for winemaking

region. The largest grape-producing countries are China, Italy, France, Spain, USA, and Türkiye.

The top US states for grape production are **California, Washington, New York**, and Texas. The USA produces 12% of the world wine in over 10,000 wineries. One acre can have 400 grape vines. While red wines are fermented in oak barrels (for 3 weeks), white wines are fermented in stainless steel tanks (for 6 weeks). Furthermore, high-quality red wines (e.g., 'Cabernet Sauvignon') benefit from aging and developing complex aromas via slow oxidation.

For more on this topic, see Morata et al. (2022) and many references therein.

8.3.5 Aroma, Flavors, and Taste

Flavor is the sensory overall sensation experience of aroma (smell) and taste to characterize food or drinks (Fig. 8.9). The main tastes are sweet, sour, salty, bitter, and umami. The main aromas are earthy, floral, fruit, spicy, and vegetal.

For more on this topic, see Cabrera et al. (2006), Carlsen et al. (2010), Balick and Cox (1996), Eichhorn et al. (2012), FAO (2017), (2024), FAOSTAT (2023), Hacisalihoglu (2011, 2021, 2022), Harshberger (1896), Judd et al. (2015), Levetin and McMahon (2020), Prance et al. (2005), Ridolfi (2019), Schmidt and Cheng (2017), Simpson and Ogorzaly (2013), Diamond (2002), Egli (2017), USDA (2023), USDA-NRCS (2000), Weber et al. (2024), Soylemezoglu et al. (2001), and many references therein.

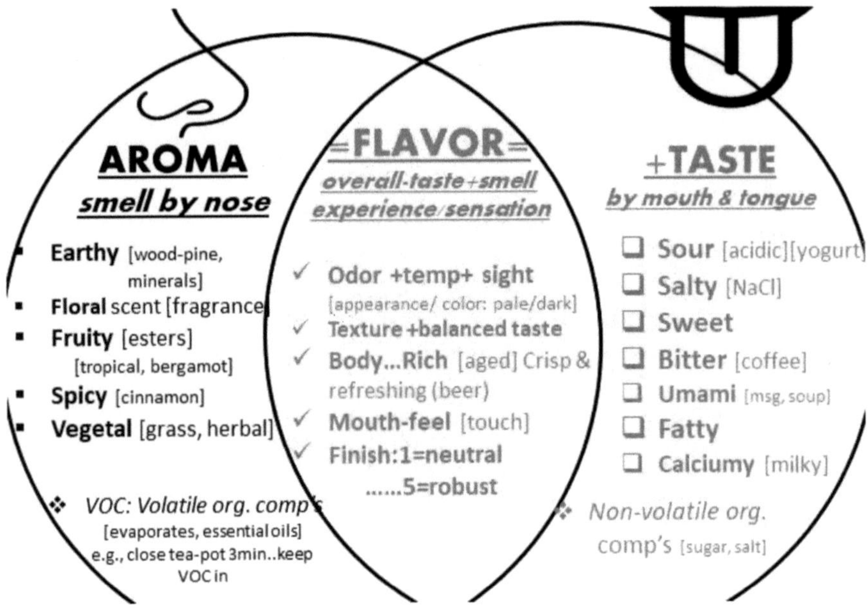

Fig. 8.9 Flavor, aroma, and taste aspects or experiences of food and drinks

8.4 Plant Profiles Bingo: How Many Can You Identify?

(P71)  *Lilium candidum*	(P72)  *Liriope muscari*
(P73)  *Lithops sp.*	(P74)  *Marchantia polymorpha*
(P75)  *Mimosa pudica*	(P76)  *Monstera deliciosa*
(P77)  *Musa acuminata*	(P78)  *Narcissus pseudonarcissus*
(P79)  *Nepenthes gracilis*	(P80)  *Nerium oleander*

8.5 Takeaway Questions

1. What is the difference between aroma, taste, and flavor?
2. Which plant species produces the true tea (black, green, oolong teas)?
3. Which plant species produces coffee?
4. Which plant species produces beer?

8.6 Now You Know the Answers

1. Flavor is the sensory overall sensation experience of aroma (smell) and taste to characterize food or drinks.
2. *Camellia sinensis.*
3. *Coffea arabica.*
4. Barley (*Hordeum vulgare*).

References

Balick MJ, Cox PA (1996) People, plants, and culture: the science of ethnobotany, New York

Bitter NQ et al (2020) Distinguishing the region-of-origin of roasted coffee beans with trace element ratios. Food Chem. https://doi.org/10.1016/j.foodchem.2020.126602

Cabrera C et al (2006) Beneficial effects of green tea—a review. J Am Coll Nutr 25:79–99

Carlsen MH et al (2010) The total antioxidant content of more than 3100 foods, beverages, spices, herbs and supplements used worldwide. Nutr J 22(9):3

Diamond J (2002) Evolution, consequences and future of plant and animal domestication. Nature 418:700–707. https://doi.org/10.1038/nature01019

Egli DB (2017) Seed biology and yield of grain crops 2e. CABI International, Boston

Eichhorn SE et al (2012) Raven biology of plants, 7th edn. Macmillan

FAO (2017) The future of food and agriculture: trends and challenges. Food and Agriculture Organization of the United Nations, Rome, Italy

FAO (2024) Food and Agriculture Organization of the United Nations. Current market situation and medium-term outlook 1 Aug 2025. http://www.fao.org

FAOSTAT (2023) FAO statistical databases. Food and Agriculture Organization of the United Nations, Rome, Italy. https://www.fao.org/faostat

Farah A (2019) Coffee: production, quality and chemistry, 1st edn. The Royal Society of Chemistry, London

Hacisalihoglu G (2011) Plant biology concepts and laboratory manual. Kendall Hunt Publishing, Iowa. ISBN-13: 9780757589133. https://www.amazon.com/Plant-Biology-Concepts-Laboratory-Manual/dp/0757589138

Hacisalihoglu, G. (2021). From growing to biology: plants 1. Florida State University. 157. ISBN13: 9781736577912

Hacisalihoglu G (2022) Level up college success: research, writing, and more key skills for STEM learners. ISBN13: 9798218066260. www.amazon.com/author/drhaci

Harshberger JW (1896) The purposes of ethnobotany. Bot Gaz 21:146–154

Illy E (2002) The complexity of coffee. Sci Am 286:86–91. https://doi.org/10.1038/scientificamerican0602-86

Judd WS et al (2015) Plant systematics: a phylogenetic approach, 4th edn. Sinauer Associates, Sunderland
Kacar B (1992) Yapraktan-bardağa çay. T.C. Ziraat Bankasi, Turkiye
Levetin E, McMahon K (2020) Plants and society, 8th edn. McGraw Hill
McKay DL, Blumberg JB (2002) The role of tea in human health: an update. J Am Coll Nutr 21:1–13
Morata A et al (2022) Grapes into wine. London, Intechopen
Morris J (2018) Coffee: a global history. Reaktion Books, London
Prance GT et al (2005) The cultural history of plants. Routledge, New York
Richelle M et al (2001) Comparison of the antioxidant activity of commonly consumed polyphenolic beverages (coffee, cocoa, and tea) prepared per cup serving. J Agric Food Chem *49*:3438–3442
Ridolfi R (2019) SDGs for people, planet, and prosperity. www.openknowledge.fao.org
Schmidt BM, Cheng DMK (2017) Ethnobotany: a phytochemical perspective. Wiley
Simpson BB, Ogorzaly MC (2013) Plants in our world: economic botany, 4th edn. McGraw-Hill
Soylemezoglu G, Agaoglu YS, Uzun HI (2001) Ampelographic characteristics and isozymic analysis of *Vitis vinifera* Spp. Sylvestris Gmel. in Southwestern Turkey. Biotechnol Biotechnol Equip *15*:106–113
USDA (2023) National Agriculture Statistics Service. https://www.nass.usda.gov/
USDA-NRCS (2000) The PLANTS database. National Resources Conservation Service. https://plants.usda.gov/
Weber C, Simnitt S, Wakefield H, Wechsler S, Swearingen B (2024) Fruit and tree nuts outlook: March 2024. United States Department of Agriculture, Washington, DC. https://www.ers.usda.gov/publications/pub-details?pubid=108880
Yazici K (2024) A'dan Z'ye Cay. Iksad Yayinevi, Ankara

Chapter 9
Fiber, Spice, and Cosmetic Plants

Fiber, Spice, and Cosmetic Plants

Contents

G. Hacisalihoglu, *Super Greens*, https://doi.org/10.1007/978-3-032-17047-7_9

9.1 Big Picture: Quick Read Chapter Infographic

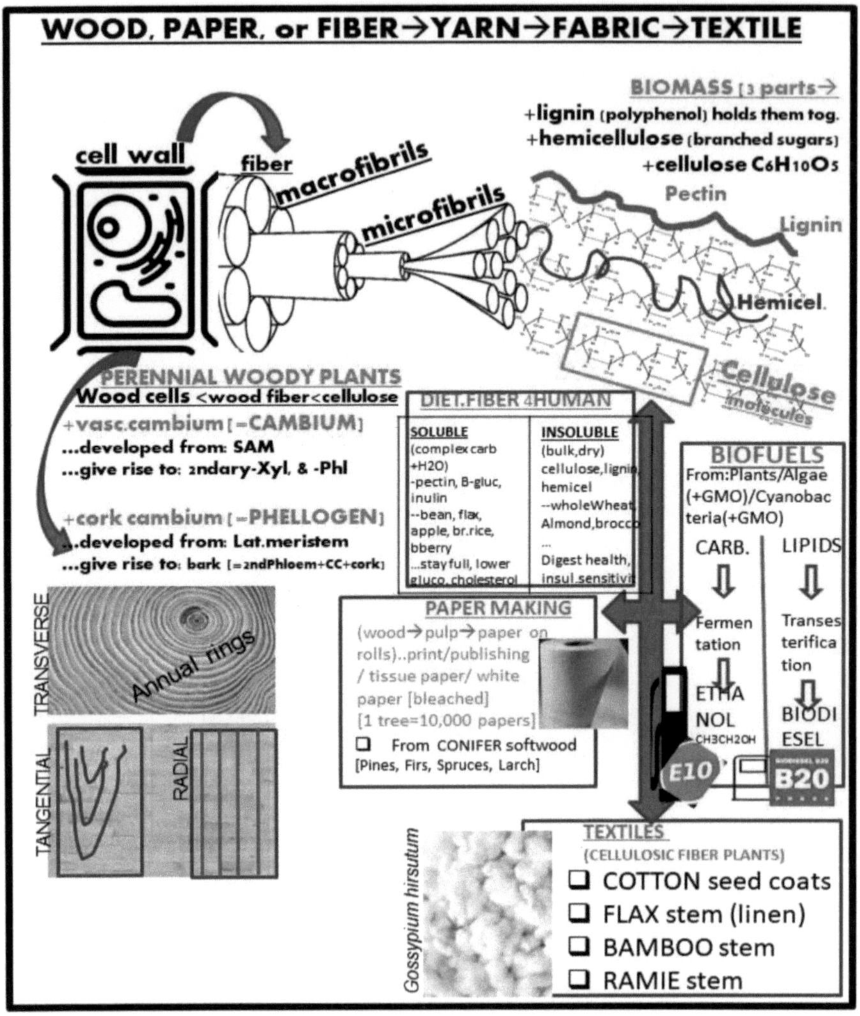

Fig. 9.1 Infographics of this chapter

9.2 Student Learning Outcomes (SLOs) of the Chapter

- Explain plant-based natural fibers and their characteristics.
- Identify cotton production worldwide.
- Distinguish between hardwoods and softwoods with examples.
- Identify fragrant plant species with scent, parts with fragrance, and growth.

9.3 Concepts of the Chapter

9.3.1 Plant-Based Natural Fibers

Plant-based natural fibers are cellulose of seed, stem, or leaf of various fiber crops (Figs. 9.2, 9.3, 9.4). Major fiber crops are cotton, hemp, jute, bamboo, flax, kenaf, and much more. Plant-based natural fibers are commonly used in clothes, textiles, ropes, cables, and many other materials (Fig. 9.2). Linen is lighter and wrinkles.

Part 1: Cotton (*Gossypium hirsutum*): Global cotton production is led by China, India, USA, and Türkiye (Fig. 9.3). Cotton is the most popular fiber crop plant in the world for its softness and moisture absorption qualities (Fig. 9.4). Turkish cotton is known for long fibers, durability, and high quality in the world textile markets.

Fig. 9.2 Fiber crops, scientific names, and major characteristics. Cotton is the most used natural textile fiber

FIBER PLANT	SCIENTIFIC NAME	CHARACTERISTICS / NATURAL PLANT-BASED
Bamboo / Jute fiber	*Bambusa vulgaris*	durable
Banana fiber	*Musa acuminata*	many uses
Coconut coir fiber	*Cocos nucifera*	many uses
Cotton fiber	*Gossypium hirsutum*	many uses
Flax / Linen fiber	*Linum usitatissimum*	many uses
Hemp fiber	*C. sativa*	many uses
Pineapple fiber	*Ananas comosus*	many uses

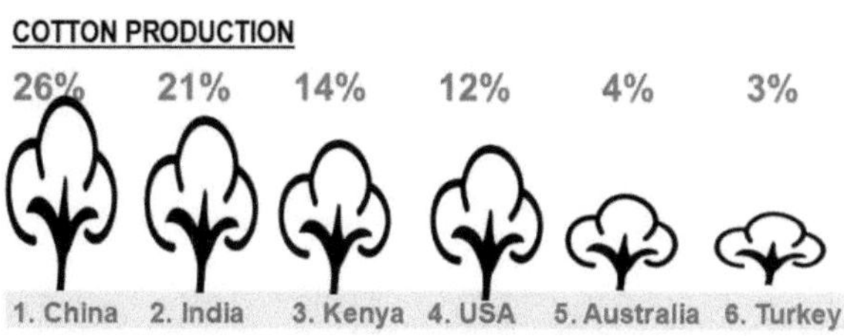

Fig. 9.3 Global cotton production with largest leading producers

Fig. 9.4 Cotton (*Gossypium hirsutum*) produces fruit (bolls) with seeds and attached hairs with fiber. (Image: G. Hacisalihoglu, PhD)

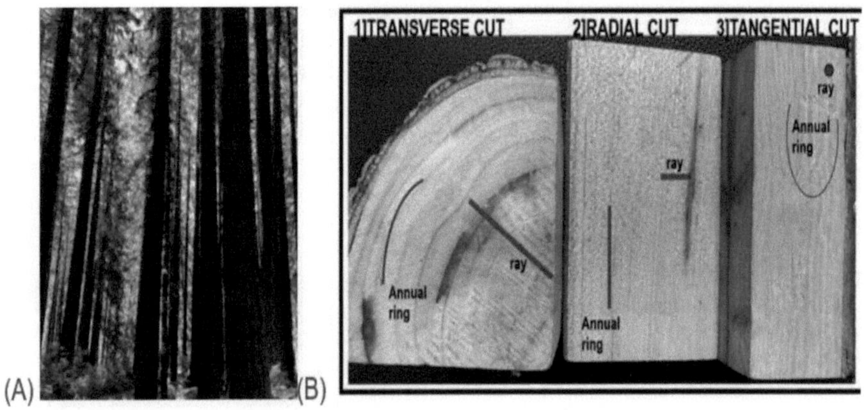

Fig. 9.5 (**a**) Redwood (*Sequoia sempervirens*) trees in California, USA; (**b**) three types of wood cuts. (1) Transverse cut; (2) radial cut; (3) tangential cut. (Image: G. Hacisalihoglu, PhD)

For more on this topic, see Bunsell (2018), Hacisalihoglu (2021), Kapadia et al. (2022), Levetin and McMahon (2020), and many references therein.

9.3.2 Hardwood and Softwood

Softwood trees grow quickly, while hardwood trees grow slowly (Figs. 9.5 and 9.6). Softwood is lighter, while **hardwood** is heavier, denser, darker color, and more expensive. The primary wood cut sections could be transverse, radial, or tangential (Fig. 9.5).

HARDWOOD SP.	Angiosperm/ deciduous	MOST COMMON IN:	**DESCRIPTION & USAGE**	**JANKA SCORE**
ROSEWOOD	*Dalbergia latifolia*	Brazil, India	Music instruments	2500
HICKORY	*Carya ovata*	USA	baseball bats, tool handles	1820
ACACIA	*Acacia auriculiformis*	Australia	brown, decorative	1750
BAMBOO	*Bambusa vulgaris*	China	Floors, furniture	1600
MAPLE	*Acer saccharum*	New York, Canada	White, music instruments [violin, cello]	1450
ASH	*Fraxinus excelsior*	USA, EU	USA native, bats, furniture	1300
OAK	*Quercus alba*	Connecticut, Illinois	Barrels, cabinets, red, white	1290
BIRCH	*Betula pendula*	N. America	white pale, doors	1260
TEAK	*Tectona grandis*	Asia	dense, brown, outdoor	1155
MANGO	*Mangifera indica*	Asia	furniture	1100
WALNUT	*Juglans regia*	USA, EU	chocolate tone, high-end furniture	1010
CHERRY	*Prunus avium*	USA	red/pink/ white, cabinets, fine furniture	955
MAHOGANY	*Swietenia macrophylla*	C. America	dense, red/brown, boats, music instruments	800
SOFTWOOD SP.	Gymnosperm/evergreen	MOST COMMON IN:	**DESCRIPTION**	**JANKA SCORE**
YEW	*Taxus baccata*	EU, Asia	hardest softwood, bows, fine furniture	1500
CEDAR	*Cedrus deodara*	USA	Light, outdoor	900
FIR	*Pseudotsuga menziesii*	Oregon, Canada	Light, doors	660
LARCH	*Larix occidentalis*	N. America	Light, decks	590
CYPRESS	*Taxodium distichum*	Louisiana	Light, outdoor, posts	510
HEMLOCK	*Conium maculatum*	USA	Light, doors	500
SPRUCE	*Picea abies*	Colorado, EU	Light, music instruments	480
REDWOOD	*Sequoia sempervirens*	California	Light, outdoor	450
PINE	*Pinus taeda*	Arkansas, EU	Light, construction, furniture	400

Fig. 9.6 Hardwoods (angiosperms, deciduous), softwoods (gymnosperms, evergreen), their scientific name, common regions, usage, description, and hardness (the higher the Janka value>1000)

Fig. 9.7 Representative softwood (pine) products, which are uniform wood with no visible pores, and hardwood (oak, poplar) products, which are complex with visible pores. (Image: G. Hacisalihoglu, PhD)

In general, hardwood comes from deciduous broad leaf trees, while softwood mainly comes from needle leaf gymnosperms (Figs. 9.6 and 9.7). Hardness of wood is measured with a Janka test. For example, Brazilian walnut has a Janka score of 3500. For more on this topic, see Wiedenhoeft (2010) and many references therein.

9.3.3 Cosmetic Plants and Essential Oils

Plants typically use **essential oils** for pollinator attraction or protection against pests. Essential oils are plant extracts from various plant parts, including flowers, fruits, leaves, or roots (Table 9.1; Fig. 9.8).

For more on this topic, see Manjari et al. (2021), Lin et al. (2018), Levetin and McMahon (2020), and many references therein.

Table 9.1 Fragrant plant species with scent, parts with fragrance, and growth

Fragrant plant	Scientific name	Fragrant part	Growth
Camellia	*Camellia lutchuensis*	Flower	Outdoor
Citrus	*Citrus* spp.	Flower	Outdoor
Daphne winter	*Daphne odora*	Flower	Outdoor
Eucalyptus	*Eucalyptus cinerea*	Leaf	Outdoor/indoor
Gardenia	*Gardenia* sp.	Flower	Outdoor/indoor
Geranium scented	*Pelargonium graveolens*	Leaf	Outdoor/indoor
Honeysuckle	*Lonicera* spp.	Flower	Outdoor
Jasmine pink	*Jasminum polyanthum*	Flower	Outdoor/indoor
Lavender	*Lavandula* spp.	Leaf	Outdoor
Lemon balm	*Melissa officinalis*	Leaf	Outdoor
Mint	*Mentha* spp.	Leaf	Outdoor/indoor
Orchid	*Cattleya* spp.	Flower	Outdoor/indoor
Peony	*Paeonia* spp.	Flower	Outdoor
Rosemary	*Rosmarinus officinalis*	Leaf	Outdoor
Roses	*Rosa* spp.	Flower	Outdoor
Sage pineap.	*Salvia elegans*	Leaf	Outdoor
Sweet olive	*Osmanthus frag*	Flower	Outdoor
Sweet pea	*Lathyrus odoratus*	Flower	Outdoor
Thyme	*Thymus serpyllum*	Leaf	Outdoor

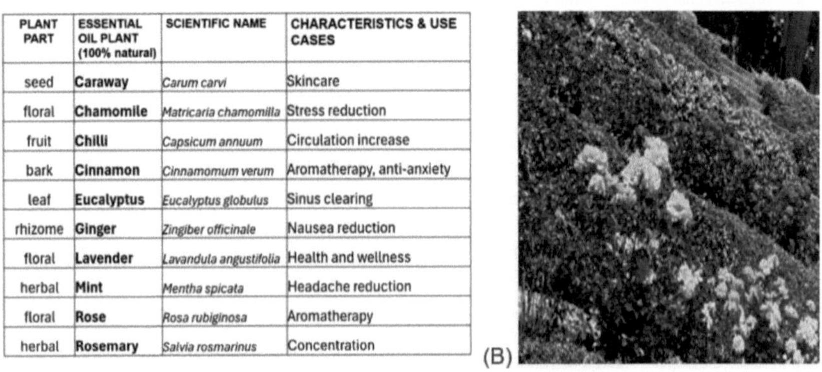

PLANT PART	ESSENTIAL OIL PLANT (100% natural)	SCIENTIFIC NAME	CHARACTERISTICS & USE CASES
seed	Caraway	*Carum carvi*	Skincare
floral	Chamomile	*Matricaria chamomilla*	Stress reduction
fruit	Chilli	*Capsicum annuum*	Circulation increase
bark	Cinnamon	*Cinnamomum verum*	Aromatherapy, anti-anxiety
leaf	Eucalyptus	*Eucalyptus globulus*	Sinus clearing
rhizome	Ginger	*Zingiber officinale*	Nausea reduction
floral	Lavender	*Lavandula angustifolia*	Health and wellness
herbal	Mint	*Mentha spicata*	Headache reduction
floral	Rose	*Rosa rubiginosa*	Aromatherapy
herbal	Rosemary	*Salvia rosmarinus*	Concentration

(A) (B)

Fig. 9.8 (**a**) Essential oil plants; (**b**) roses. (Image: G. Hacisalihoglu, PhD)

9.3.4 Spices and Herbs

Spices and herbs have been in use for thousands of years globally (Fig. 9.9a). Spices and herbs are used for adding flavor to food as well as food preservatives (Fig. 9.9b). The top herbs are oregano, mint, rosemary, cilantro, basil, thyme, sage, chives, dill, and parsley.

For more on this topic, see Peter 2006 and many references therein.

Part 2: Hot Peppers: Scoville Heat Units (SHU) measures the heat or spiciness levels. While 0 HSU is sweet (e.g., bell pepper, shishito), 1 million SHU (e.g., 'Carolina Reaper') is super-hot. Full ripening will also increase the heat levels (Fig. 9.10). Hot peppers contain capsaicin metabolites that give them the heat and burning sensation in the mouth.

9.3.5 Plant Natural Dyes

Plant dyes are their natural pigments obtained from plant organs such as root, leaf, flower, or fruits. There are numerous plant species with natural plant dyes, including indigo, marigold, and madder (Fig. 9.11). For example, madder (*Rubia tinctoria*) roots contain red dye. Henna (*Lawsonia inermis*) leaves contain reddish brown color that is commonly used as natural hair dye worldwide. Henna has been used in some regions for thousands of years for ceremonial purposes.

For more on this topic, see Bunsell (2018), Diamond (2002), Balick and Cox (1996), Eichhorn et al. (2012), Reece et al. (2014), FAO (2017, 2024), FAOSTAT 2023, Hacisalihoglu (2011, 2021, 2022), Harshberger (1896), Judd et al. (2015), Levetin and McMahon (2020), Prance et al. (2005), Ridolfi (2019), Schmidt and Cheng (2017), Simpson and Ogorzaly (2013), USDA (2023), USDA-NRCS (2000), Weber et al. (2024), and many references therein.

(A)

SPICE PLANT (flower, fruit, seed, bark, root)	SCIENTIFIC NAME	CHARACTERISTICS
Black pepper	*Piper sp.*	Spice
Cardamom	*Elettaria cardamomum*	Spice
Chili pepper	*Capsicum annum*	Spice
Cinnamon	*Cinnamomum verum*	Drinks, anti-diabetic
Cumin	*Cuminum cyminum*	Curries, cancer fighting
Garlic powder	*Allium sativum*	Spice
Ginger	*Zingiber officinale*	Tea, digestion
Nutmeg	*Myristica fragrans*	Spice
Oregano	*Origanum vulgare*	Spice
Turmeric	*Curcuma longa*	Pain reduce, heart healthy

(B)

Fig. 9.9 (**a**) Common spices and characteristics; (**b**) spices commonly found in stores. (Image: G. Hacisalihoglu, PhD)

Capsicum species	Scoville Heat Units (SHU) For [CAPSAICIN] [capsaicin→16 Mil.SHU]
Pepper X (World's hottest Pepper)	2.7 Mil (super hot)
Carolina Reaper	2.2 Mil (super hot)
Trinidad Scorpion	2 Mil (super hot)
Naga Viper	1.3 Mil (super hot)
Ghost Pepper	1 Mil (super hot)
Red Savina Habanero	0.5 Mil (super hot)
Orange Habanero	0.3 Mil (extremely hot)
Cayenne	50 K (very hot)
Tabasco	50 K (very hot)
Serrano	25 K (very hot)
Jalapeno	8 K (hot)
Urfa	7.5 K (hot)
Anaheim/ Banana Pepper	2 K (mild hot)
Bell Pepper	0 (not hot / sweet)

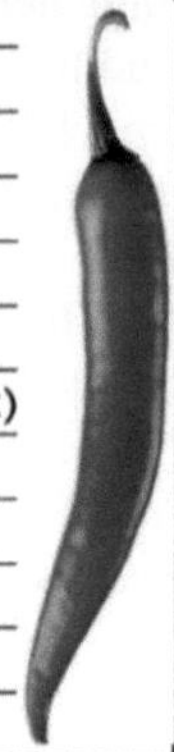

Fig. 9.10 Capsaicin compound concentration levels of chili peppers as measured with Scoville Heat Units (SHU) of various hot peppers (*Capsicum* spp.). Chili peppers are typically bred for higher spiciness such as 'Carolina Reaper' cultivar

PLANT DYES	SCIENTIFIC NAME	CHARACTERISTICS / NATURAL PLANT-BASED
Henna	*Lawsonia inermis*	red dye
Indigo	*Indigofera tinctoria*	blue dye
Madder	*Rubia tinctoria*	Red dye
Marigold	*Tagetes patula*	gold dye
Tea leaf	*Camelia sinensis*	Brown dye
Turmeric	*Curcuma longa*	yellow dye
Weld	*Reseda luteola*	Bright yellow

Fig. 9.11 Plant natural dyes and their characteristics. The oldest known natural red dye comes from the roots of common madder plants

9.4 Plant Profiles Bingo: How Many Can You Identify?

(P81)  *Ocimum basilicum*	(P82)  *Olea europaea*
(P83)  *Oryza sativa*	(P84)  *Pachira aquatica*
(P85)  *Pachycereus schottii*	(P86)  *Paeonia tenuifolia*
(P87)  *Pennisetum setaceum*	(P88)  *Petroselinum crispum*
(P89)  *Phalaenopsis aphrodite*	(P90)  *Philodendron bipinnatifidum*

9.5 Takeaway Questions

1. Which plant species provides the textile linen?
2. Which plant species provides the blue textile color?
3. Which plant species provides the essential oil for Earl Grey tea in black tea?
4. Which plant species provides the soft, most widely used textile material?

9.6 Now You Know the Answers

1. Flax (*Linum usitatissimum*)
2. Indigo (*Indigofera tinctoria*)
3. Bergamot orange (*Citrus bergamia*)
4. Cotton (*Gossypium hirsutum*)

References

Balick MJ, Cox PA (1996) People, plants, and culture: the science of ethnobotany, New York

Bunsell AR (2018) Introduction to the science of fibers. In: Handbook of properties of textile and technical fibres, 2nd edn. Woodhead Publishing, pp 1–20

Diamond J (2002) Evolution, consequences and future of plant and animal domestication. Nature 418:700–707. https://doi.org/10.1038/nature01019

Eichhorn SE et al (2012) Raven biology of plants, 7th edn. Macmillan

FAO (2017) The future of food and agriculture: trends and challenges. Food and Agriculture Organization of the United Nations, Rome

FAO (2024) Food and Agriculture Organization of the United Nations. Current market situation and medium-term outlook 1 Aug 2025. http://www.fao.org

FAOSTAT (2023) FAO statistical databases. Food and Agriculture Organization of the United Nations, Rome. https://www.fao.org/faostat

Hacisalihoglu G (2011) Plant biology concepts and laboratory manual. Kendall Hunt Publishing, Iowa. ISBN-13: 9780757589133. https://www.amazon.com/Plant-Biology-Concepts-Laboratory-Manual/dp/0757589138

Hacisalihoglu G (2021) From growing to biology: plants, 1st edn. Florida State University, p 157. ISBN13: 9781736577912

Hacisalihoglu G (2022) Level up college success: research, writing, and more key skills for STEM learners. ISBN13: 9798218066260. www.amazon.com/author/drhaci

Harshberger JW (1896) The purposes of ethnobotany. Bot Gaz 21:146–154

Judd WS et al (2015) Plant systematics: a phylogenetic approach, 4th edn. Sinauer Associates, Sunderland

Kapadia P et al (2022) Extraction of high-value chemicals from plants for technical and medical applications. Int J Mol Sci 23:10334

Levetin E, McMahon K (2020) Plants and society, 8th edn. McGraw Hill

Lin TK et al (2018) Anti-inflammatory and skin barrier repair effects of topical application of some plant oils. Int J Mol Sci 19:70

Manjari KS et al (2021) Biodiversity and importance of plant bioprospecting in cosmetics. In: Bioprospecting of plant biodiversity for industrial molecules. Wiley, pp 189–210

Peter KV (2006) Handbook of herbs and spices. Woodhead Publishing Limited, Cambridge

Prance GT et al (2005) The cultural history of plants. Routledge, New York
Reece JB et al (2014) Campbell biology, 10th edn. Pearson, Boston
Ridolfi R (2019) SDGs for people, planet, and prosperity. www.openknowledge.fao.org
Schmidt BM, Cheng DMK (2017) Ethnobotany: a phytochemical perspective. Wiley
Simpson BB, Ogorzaly MC (2013) Plants in our world: economic botany, 4th edn. McGraw-Hill
USDA (2023) National Agriculture Statistics Service. https://www.nass.usda.gov/
USDA-NRCS (2000) The PLANTS database. National Resources Conservation Service. https://plants.usda.gov/
Weber C, Simnitt S, Wakefield H, Wechsler S, Swearingen B (2024) Fruit and tree nuts outlook: march 2024. United States Department of Agriculture, Washington, DC. https://www.ers.usda.gov/publications/pub-details?pubid=108880
Wiedenhoeft A (2010) Structure and function of wood. In: Wood handbook: wood as an engineering material. Forest Products Laboratory, Forest Service, U.S. Department of Agriculture, Madison

Chapter 10
Medicinal Plants

Medicinal Plants

Contents

G. Hacisalihoglu, *Super Greens*, https://doi.org/10.1007/978-3-032-17047-7_10

10.1 Big Picture: Quick Read Chapter Infographic

10.2 Student Learning Outcomes (SLOs) of the Chapter

- Explain plant secondary metabolites and their importance for plants and humans.
- Identify various plant secondary chemicals in different plant species.
- Distinguish between primary and secondary metabolic pathways.
- Identify plant species for malaria, anticancer, antidiabetic treatment purposes.

10.3 Concepts of the Chapter

10.3.1 Plant Metabolic Pathways

Plant metabolic pathways are primary pathways and secondary pathways. Primary metabolites are used in growth and reproduction, including lipids, sugars, and amino acids. Secondary metabolites are used for defense and hormone functions, including phenolics, terpenes, nitrogen compounds, and sulfur compounds (Fig. 10.2).

For more on this topic, see Schwender (2009) and many references therein.

10.3.2 Medicinal Plants: Focus on Plant Secondary Compounds

Medicinal plants have been used for therapeutic purposes for millenniums. Plants generate multiple diverse metabolites with great potential in medicine and other health related fields (Table 10.1; Fig. 10.1). Potential target plant secondary compounds are often in trace concentrations. By understanding the regulation of secondary metabolites, scientists can use these small quantities of compounds for treatment of diseases, including diabetics, cancer, malaria, and much more (Fig. 10.1). Furthermore, how secondary metabolite compounds interact with enzymes and nutrients is also of interest in molecular biology and bioinformatics. Caution: All medicinal plant information in this chapter is for educational and reference purposes only, and consulting a healthcare professional is recommended for all medicinal or health uses.

For more on this topic, see Berg et al. ((2019)), Balunas and Kinghorn (2005), Amos et al. (1997), Aggarwal (2007), Diamond (2002), Gurib-Fakim (2006), Gurdal and Kultur (2013), Dragland et al. (2003), Munos-Acevedo et al. (2018), Longaray

PLANT SECONDARY CHEMICALS, THEIR ACTIVE COMPOUNDS, & FOUND IN

PHENOLICS	TERPENES	N-COMPS	S-COMPS
Alkaloids	**Brassionoids**	**Cyanogenic glycosides**	**Defensins**
~Artemisia leaf	**Carotene**	~Cassava	
(malaria)	**Caryophyllene**	~Oleander leaf	
Flavanoids	~black-pepper	~Aloe	**Disulphides**
~Cinnamon zeylanicum bark	(anti-inflam.)	~Milkweed	~Onion
~Quercetin, Hesperidin	~parsley	~Seeds of apple	~Garlic
~Naringin, Rutin	**Gibberellins**	cherry,pears	(anti-viral)
~Rutin (diabetes)	**Limonene**	peach,plum	
~Caffeine, Nicotine, Morp	~Citrus, Bergamot		**Glucosinolates**
~Saw palmetto (prostate)	~Carrot	**Taxus**	~Brassicas
~Guanidine (t2d)	(anti-anxiety)		
^^Galega officinalis (METF)			**Phytochelatins**
Flavonoglycosides	**Linalool**		
~Ginkgo leaf	~Lavender		
(alzheimer)	(anti-anxiety)		
Isoflavonoids	**Myrcene**		
Lignins	~hops,ginger		
~Flax seed	~cardamom		
(cholesterol)	~Lemongrass		
Phenols	(anti-bac/fungal)		
Polyphenols	**Pinene [anti-anx.]**		
~Ginger	~Eucalyptus		
(anti-inflammatory)	~Pine needles		
Salicyclic acid	~Pepper		
~Willow bark	~Oregano		
(Aspirin)	~Basil,dill		
Saponins			
~Berberis bark (t2d)			
~Aloe vera (skin)			
Sterols			
Tannins			
~black-tea (t2d)			
~witch-hazel, Grapes			

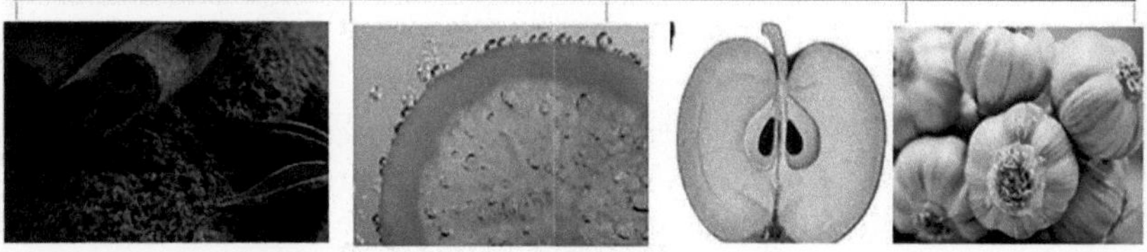

Fig. 10.1 Infographics of this chapter

et al. (2007), Lattanzio et al. (2008), Johnson (2004), Holton et al. (1995), Levetin and McMahon (2020), and many references therein.

What are the top medicinal plants which are most recommended to use for burns, digestion, headaches, relaxing, calming, inflammation, stress reduction, antimicrobial, antifungal, and health benefits?

MAJOR PLANT METABOLIC PATHWAYS & METABOLITE PRODUCTS

	PATHWAY	TYPE	SITE	REACTANT	PRODUCT/metabolite
1° PATHWAYS functions: growth, reproduction	PHOTOSYNTHESIS [Calvin Cycle]	1° Metabolism	Chloroplast	CO2, RuBP	Sugars (glucose, sucrose)
	PLANT RESPIRATION [Glycolysis]	1° Metabolism	Cytoplasm	Glucose	Pyruvate
	PLANT RESPIRATION [Citric Acid Cycle/ TCA]	1° Metabolism	Mitochondria	Acetyl CoA	ATP, NADH, FADH2
	AMINO ACID METABOLISM	1° Metabolism	Chloroplast, Cytoplasm	Org. acids (α-ketoglutarate)	Amino acids (Glutamate, Aspartate)
	LIPID METABOLISM	1° Metabolism	ER, Cytoplasm	Fatty acids, glycerol	Triacyclglycerols, Phospholipids
2° PATHWAYS functions: defense, hormones, pigments	TERPENOID PATHWAY	2° Metabolism	Cytoplasm	Pyruvate	Isoprenoids, Terpenes
	ALKALOID PATHWAY	2° Metabolism	Cytoplasm	Amino acids (tryptophan)	Alkaloids (nicotine)
	GLUCOSINOLATE PATHWAY	2° Metabolism	Cytoplasm	Methionine Tryptophan	Glucosinolates
	MEVALONATE PATHWAY	2° Metabolism	Cytoplasm	Acetyl CoA	Terpenes
	POLYPHENOL BIOSYNTHESIS	2° Metabolism	Cytoplasm	Phenylalanine	Polyphenols (Anthocyanins)

Fig. 10.2 Representative ten major plant metabolic pathways. Plant metabolic pathways are primary pathways for development (amino acids, sugars, lipids, and more) and secondary pathways are pathways for environmental interactions (phenols, flavonoids, terpenes, etc.)

Part 1: Aloe: Aloe (*Aloe vera*) is a medicinal plant species which is traditionally renowned for supporting skin care, digestion, sugar control, and anti-inflammatory (Fig. 10.3a). Disclaimer: It must be discontinued in case of side effects, allergies, irritation, or rash sensitivity situations, if any.

Part 2: Calendula: Calendula (*Calendula officinalis*) is a medicinal plant species which is traditionally renowned for supporting skin-related issues, immunity, anti-inflammatory, and digestive issues (Fig. 10.3b). Disclaimer: It must be discontinued in case of side effects, allergies, irritation, or rash sensitivity situations, if any.

Part 3: Chamomile: Chamomile (*Matricaria chamomilla*) is a medicinal plant species which is traditionally renowned for supporting antianxiety, digestive health, anti-inflammatory, and antioxidant (Fig. 10.3c). Most common usage of this plant is

Table 10.1 Medicinal plants, characteristics, and the compound used in medicine. All medicinal plant information in this chapter is for educational and reference purposes only, and consulting a healthcare professional is recommended for all medicinal or health uses

Medical potential	Medicinal plant	Part used	Scientific name	Compound
Antidiabetic	Cinnamon	Bark	*Cinnamomum zeylanicum*	Cinnamaldehyde
	French lilac, goat's rue	Flower, leaf, stem	*Galega officinalis*	Metformin (patented)
	Neem	Leaf, seed	*Azadirachta indica*	Nimbin
	Bitter melon	Fruit	*Mommordica charantia*	Triterpenoid
	Fenugreek	Seed	*Trigonella foenum*	Saponin
	Gooseberry	Fruit	*Phyllanthus emblica*	Tannins
	Java plum	Seed	*Syzygium cumini*	Polyphenols
Anticancer	Turmeric	Rhizome	*Curcuma longa*	Curcumin
	Pacific yew	Bark	*Taxus brevifolia*	Taxol (patented)
	Green tea	Leaf	*Camelia sinensis*	Epigallocatechin
	Ginseng	Root	*Panax ginseng*	Ginsenosides
Anti-inflammatory	Jergon Sacha	Root	*Dracontium loretense*	Phenols
Antiviral	Oregano	Oil	*Origanum vulgare*	Carvacrol
	Ginger	Rhizome	*Zingiber officinale*	Gingerone
	Cats claw	Bark	*Uncaria tomentosa*	Phytochemicals
	Garlic	Bulb	*Allium sativum*	Allicin
Antibacterial/ cold	Elderberry	Berry	*Sambucus nigra*	Polyphenols
	Peppermint	Essential oil	*Mentha piperita*	Menthol
	Yarrow	Essential oil	*Achillea millefolium*	Terpenes
	Coneflower	Essential oil	*Echinacea purpurea*	Various
	Sunflower	Leaf	*Helianthus* spp.	Various
Headache relief	Lavender	Essential oil	*Lavandula angustifolia*	Linalool
	Willow	Bark	*Salix alba*	Salicin
Cuts/ wound healing	Aloe vera	Leaf gel	*Aloe barbadensis*	Aloin
Malaria treatment	Quinine	Bark	*Cinchona calisaya*	Quinine
Skin irritation	Witch hazel	Leaf /bark	*Hamamelis virginiana*	Tannins
	Calendula	Flower	*Calendula officinalis*	Triterpenoids
Diarrhea	Blackberry	Root	*Rubus fruticosus*	Tannins
	Guava	Leaf	*Psidium guajava*	Flavonoids
Constipation	Senna	Leaf	*Senna alexandriana*	Sennosides

Fig. 10.3 Medicinal plant species with various health benefits. (**a**) Aloe; (**b**) calendula; (**c**) chamomile; (**d**) cinnamon. (Image: G. Hacisalihoglu, PhD)

as tea. Disclaimer: It must be discontinued in case of side effects, allergies, irritation, or rash sensitivity situations, if any.

Part 4: Cinnamon: Cinnamon (*Cinnamomum verum*) is a medicinal plant species which is traditionally renowned for supporting antioxidant, anti-inflammatory, digestion treatments, anticancer, and sugar control in the blood (Fig. 10.3d). Disclaimer: It must be discontinued in case of side effects, allergies, irritation, or rash sensitivity situations, if any.

For more on this topic, see Gurib-Fakim (2006), Gurdal and Kultur (2013), Dragland et al. (2003), Munos-Acevedo et al. (2018), Longaray et al. (2007), Lattanzio et al. (2008), Johnson (2004), Holton et al. (1995), Levetin and McMahon (2020), and many references therein.

Part 5: Dandelion: Dandelion (*Taraxacum officinale*) is a medicinal plant species which is traditionally renowned for supporting digestion, sugar control, antioxidant, and anti-inflammatory properties (Fig. 10.4a). Disclaimer: It must be discontinued in case of side effects, allergies, irritation, or rash sensitivity situations, if any.

Part 6: Echinacea: Echinacea (purple coneflower, *Echinacea purpurea*) is a medicinal plant species which is traditionally renowned for supporting cold treatments, anti-inflammatory, anxiety reduction, and skin care treatments (Fig. 10.4b). Most common usage of this plant is making tea with its dried flowers. Disclaimer: It

Fig. 10.4 (**a**) Medicinal plant species with various health benefits. Dandelion; (**b**) echinacea; (**c**) elderberry; (**d**) fennel. (Image: G. Hacisalihoglu, PhD)

must be discontinued in case of side effects, allergies, irritation, or rash sensitivity situations, if any.

Part 7: Elderberry: Elderberry (*Sambucus canadensis*) is a medicinal plant species which is traditionally renowned for supporting antioxidant, heart health, and anti-inflammatory properties (Fig. 10.4c). Disclaimer: It must be discontinued in case of side effects, allergies, irritation, or rash sensitivity situations, if any.

Part 8: Fennel: Fennel (*Foeniculum vulgare*) is a medicinal plant species which is traditionally renowned for supporting hormones, antioxidant, anti-inflammatory, and heart health (Fig. 10.4d). Disclaimer: It must be discontinued in case of side effects, allergies, irritation, or rash sensitivity situations, if any.

Part 9: Garlic: Garlic (*Allium sativum*) is a medicinal plant species which is traditionally renowned for supporting antimicrobial, heart health, anti-inflammatory, and cholesterol lowering treatments (Fig. 10.5a). Disclaimer: It must be discontinued in case of side effects, allergies, irritation, or rash sensitivity situations, if any.

Part 10: Ginger: Ginger (*Zingiber officinale*) is a medicinal plant species which is traditionally renowned for supporting digestive treatments, nausea issues, anti-inflammatory, and antimicrobial treatments (Fig. 10.5b). Disclaimer: It must be discontinued in case of side effects, allergies, irritation, or rash sensitivity situations, if any.

Part 11: Lavender: Lavender (*Lavandula angustifolia*) is a medicinal plant species which is traditionally renowned for supporting antimicrobial, anxiety

Fig. 10.5 Medicinal plant species with various health benefits. (**a**) Garlic; (**b**) ginger; (**c**) lavender; (**d**) lemon balm. (Image: G. Hacisalihoglu, PhD)

reduction, anti-inflammatory, and antimicrobial treatments (Fig. 10.5c). Disclaimer: It must be discontinued in case of side effects, allergies, irritation, or rash sensitivity situations, if any.

Part 12: Lemon balm: Lemon balm (*Melissa officinalis*) is a medicinal plant species which is traditionally renowned for supporting calming, anxiety reduction, heart health, digestion, and sleep improvement (Fig. 10.5d). Most common usage of this plant is making tea with its fresh leaves. Disclaimer: It must be discontinued in case of side effects, allergies, irritation, or rash sensitivity situations, if any.

For more on this topic, see Munos-Acevedo et al. (2018), Longaray et al. (2007), Lattanzio et al. (2008), Johnson (2004), Holton et al. (1995), Levetin and McMahon (2020), and many references therein.

Part 13: Rosemary: Rosemary (*Salvia rosmarinus*) is a medicinal plant species which is traditionally renowned for supporting anti-inflammation, blood circulation, cognition health, and antioxidant (Fig. 10.6a). Most common usage of this plant is making tea with its dried leaves. Disclaimer: It must be discontinued in case of side effects, allergies, irritation, or rash sensitivity situations, if any.

Part 14: Sage: Sage (*Salvia officinalis*) is a medicinal plant species which is traditionally renowned for supporting hormonal balance, cough, oral health, sugar levels, cholesterol levels, antimicrobial, and antioxidant (Fig. 10.6b). Most common usage of this plant is making tea with its fresh or dried leaves. Disclaimer: It must be

Fig. 10.6 Medicinal plant species with various health benefits. (**a**) Rosemary; (**b**) sage; (**c**) St. John's wort; (**d**) thyme. (Image: G. Hacisalihoglu, PhD)

discontinued in case of side effects, allergies, irritation, or rash sensitivity situations, if any.

Part 15: St. John's Wort: St. John's wort (*Hypericum perforatum*) is a medicinal plant species which is traditionally renowned for supporting depression treatment, digestive issues treatment, cholesterol, and sugar level management (Fig. 10.6c). Most common usage of this plant is capsules or oil. Disclaimer: It must be discontinued in case of side effects, allergies, irritation, or rash sensitivity situations, if any.

Part 16: Thyme: Thyme (*Thymus vulgaris*) is a medicinal plant species which is traditionally renowned for supporting respiration, skin conditions, as well as antioxidant (Fig. 10.6d). Disclaimer: It must be discontinued in case of side effects, allergies, irritation, or rash sensitivity situations, if any.

Part 17: Ginkgo: Ginkgo (*Ginkgo biloba*) is a medicinal plant species which is traditionally renowned for supporting memory improvement benefits. For example, Ginkgo is commonly used in dementia treatment (Fig. 10.7). It must be discontinued in case of side effects, allergies, irritation, or rash sensitivity situations, if any.

Part 18: Goat's rue: Goat's rue (*Galega officinalis*) is a medicinal plant species which is traditionally renowned for supporting diabetes management and lowering blood sugar. It is used in popular diabetes medicine named metformin (Fig. 10.8).

Fig. 10.7 Medicinal plant species of Ginkgo with various health benefits

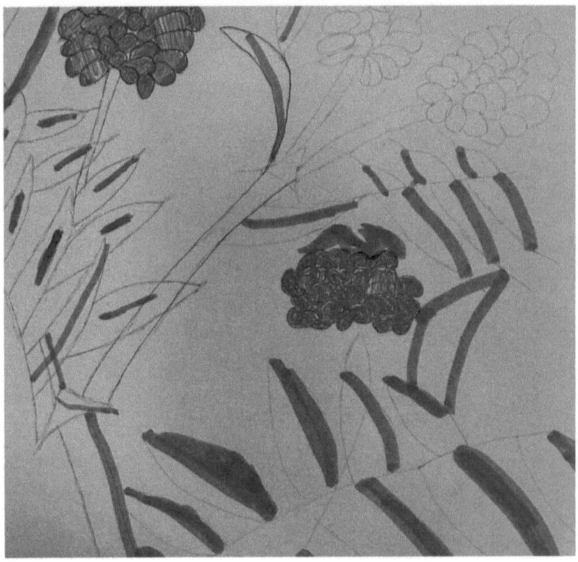

Fig. 10.8 Medicinal plant species of goat's rue with various health benefits. The metabolite guanidine has been used in metformin medication [synthesized in 1922 and used for diabetes treatment since 1950] to lower insulin resistance and help controlling type 2 diabetes sugar levels

Disclaimer: It must be discontinued in case of side effects, allergies, irritation, or rash sensitivity situations, if any.

For more on this topic, see Tyler and Robbers (1988), Serrano et al. (2018), Rehecho et al. (2011), Passalacqua et al. (2007), Palmese et al. (2001), Odot (2004), Balick and Cox (1996), Eichhorn et al. (2012), Reece et al. (2014), FAO (2017, 2024), FAOSTAT (2023), Hacisalihoglu (2011, 2021, 2022), Harshberger (1896), Judd et al. (2015), Levetin and McMahon (2020), Prance et al. (2005), Ridolfi (2019), Schmidt and Cheng (2017), Simpson and Ogorzaly (2013), USDA (2023), USDA-NRCS (2000), Weber et al. (2024), and many references therein.

10.4 Plant Profiles Bingo: How Many Can You Identify?

(P91)  *Prunus serrullata*	(P92)  *Psilotum nudum*
(P93)  *Ranunculus asiaticus*	(P94)  *Rhododendron fashion*
(P95)  *Rosa chryslerimperial*	(P96)  *Serenoa repens*
(P97)  *Saccharum officinarum*	(P98)  *Salvia splendens*
(P99)  Eichhorn, S. E. et al. (2013). *Raven Biology of Plants 7e*. Macmillan.*Sansevieria trifasciata*	(P100)  *Sarracenia purpurea*

10.5 Takeaway Questions

1. What are the health benefits of medicinal plants?
2. TRUE/FALSE: Garlic has been shown to have anti-inflammatory properties.
3. TRUE/FALSE: Chamomile has been shown to have anxiety lowering properties.
4. TRUE/FALSE: St. John's wort has been shown to antidepression properties.

10.6 Now You Know the Answers

1. Many medicinal plant species promote various health benefits from anxiety reduction to antidiabetic to anticancer properties.
2. True.
3. True.
4. True.

References

Aggarwal BB (2007) The molecular targets and therapeutic uses of curcumin in health and disease. Adv Exp Med Biol 595

Amos A, McCarthy D, Zimmet P (1997) The rising global burden of diabetes and its complications: estimates and projections to the year 2010. Diabetic Med 14:S1–S85

Balick MJ, Cox PA (1996) People, plants, and culture: the science of ethnobotany, New York

Balunas MJ, Kinghorn AD (2005) Drug discovery from medicinal plants. Life Sci 78:431–441

Berg JM, Tymoczko JL, Gatto GJ, Stryer L (2019) Biochemistry, 9th edn. W.H. Freeman/Macmillan Learning, New York

Diamond J (2002) Evolution, consequences and future of plant and animal domestication. Nature 418:700–707. https://doi.org/10.1038/nature01019

Dragland S et al (2003) Several culinary and medicinal herbs are important sources of dietary antioxidants. J Nutr 133:1286–1290

Eichhorn SE et al (2012) Raven biology of plants, 7th edn. Macmillan

FAO (2017) The future of food and agriculture: trends and challenges. Food and Agriculture Organization of the United Nations, Rome, Italy

FAO (2024) Food and Agriculture Organization of the United Nations. Current market situation and medium-term outlook 1 Aug 2025. http://www.fao.org

FAOSTAT (2023) FAO statistical databases. Food and Agriculture Organization of the United Nations, Rome. https://www.fao.org/faostat

Gurdal B, Kultur S (2013) An ethnobotanical study of medicinal plants in Marmaris (Muğla, Turkey). J Ethnopharmacol 146:113–126

Gurib-Fakim A (2006) Medicinal plants: traditions of yesterday and drugs of tomorrow. Mol Asp Med 27:1–93

Hacisalihoglu G (2011) Plant biology concepts and laboratory manual. Kendall Hunt Publishing, Iowa. ISBN-13: 9780757589133. https://www.amazon.com/Plant-Biology-Concepts-Laboratory-Manual/dp/0757589138

Hacisalihoglu, G. (2021). From growing to biology: plants 1 Florida State University. 157. ISBN13: 9781736577912

Hacisalihoglu G (2022) Level up college success: research, writing, and more key skills for STEM learners. ISBN13: 9798218066260. www.amazon.com/author/drhaci
Harshberger JW (1896) The purposes of ethnobotany. Bot Gaz 21:146–154
Holton RA et al (1995) Taxane anticancer agents: basic science and current status, ACS symposium series no. 583. American Chemical Society, Washington, DC, pp 288–301
Johnson IT (2004) New approaches to the role of diet in the prevention of cancers of the alimentary tract. Mutat Res 551:9–28. https://doi.org/10.1016/j.mrfmmm.2004.02.017
Judd WS et al (2015) Plant systematics: a phylogenetic approach, 4th edn. Sinauer Associates, Sunderland
Lattanzio P et al (2008) Plant phenolics-secondary metabolites with diverse functions. In: Recent advances in polyphenol research. Wiley, pp 1–35
Levetin E, McMahon K (2020) Plants and society, 8th edn. McGraw Hill
Longaray D et al (2007) Antibacterial activity of the essential oils of Salvia officinalis L. and Salvia triloba L. cultivated in South Brazil. Food Chem 100:603–608
Munos-Acevedo A et al (2018) Some medicinal and aromatic plants used for the treatment of cancer/tumors, with special reference to its scientific validations. In: Ethnobotany. CRS Press, p 32
Odot J (2004) In vitro and in vivo anti-tumoral effect of curcumin against melanoma cells. Int J Cancer 111:381–387
Palmese MT et al (2001) An ethno-pharmacobotanical survey in the Sarrabus district (south-East Sardinia). Fitoterapia 72:619–643
Passalacqua NG et al (2007) Contribution to the knowledge of the folk plant medicine in Calabria region (southern Italy). Fitoterapia 78:52–68
Prance GT et al (2005) The cultural history of plants. Routledge, New York
Reece JB et al (2014) Campbell biology, 10th edn. Pearson, Boston
Rehecho S et al (2011) Ethnopharmacological survey of medicinal plants in Nor-Yauyos, a part of the Landscape Reserve Nor-Yauyos-Cochas, Peru. J Ethnopharmacol 133:75–85
Ridolfi R (2019) SDGs for people, planet, and prosperity. www.openknowledge.fao.org
Schmidt BM, Cheng DMK (2017) Ethnobotany: a phytochemical perspective. Wiley
Schwender J (2009) Plant metabolic networks. Springer, Dordrecht
Serrano A et al (2018) Bioactive compounds and extracts from traditional herbs and their potential anti-inflammatory health effects. Medicine (Baltimore) 5:76
Simpson BB, Ogorzaly MC (2013) Plants in our world: economic botany, 4th edn. McGraw-Hill
Tyler VE, Robbers JE (1988) Pharmacognosy, 9th edn. Lea & Febiger, Philadelphia
USDA (2023) National Agriculture Statistics Service. https://www.nass.usda.gov/
USDA-NRCS (2000) The PLANTS database. National Resources Conservation Service. https://plants.usda.gov/
Weber C, Simnitt S, Wakefield H, Wechsler S, Swearingen B (2024) Fruit and tree nuts outlook: march 2024. United States Department of Agriculture, Washington, DC. https://www.ers.usda.gov/publications/pub-details?pubid=108880

Chapter 11
Diets, Nutrition, Superfoods, and Super Agers

Diets, Nutrition, Superfoods, and Super Agers

Contents

G. Hacisalihoglu, *Super Greens*, https://doi.org/10.1007/978-3-032-17047-7_11

11.1 Big Picture: Quick Read Chapter Infographic

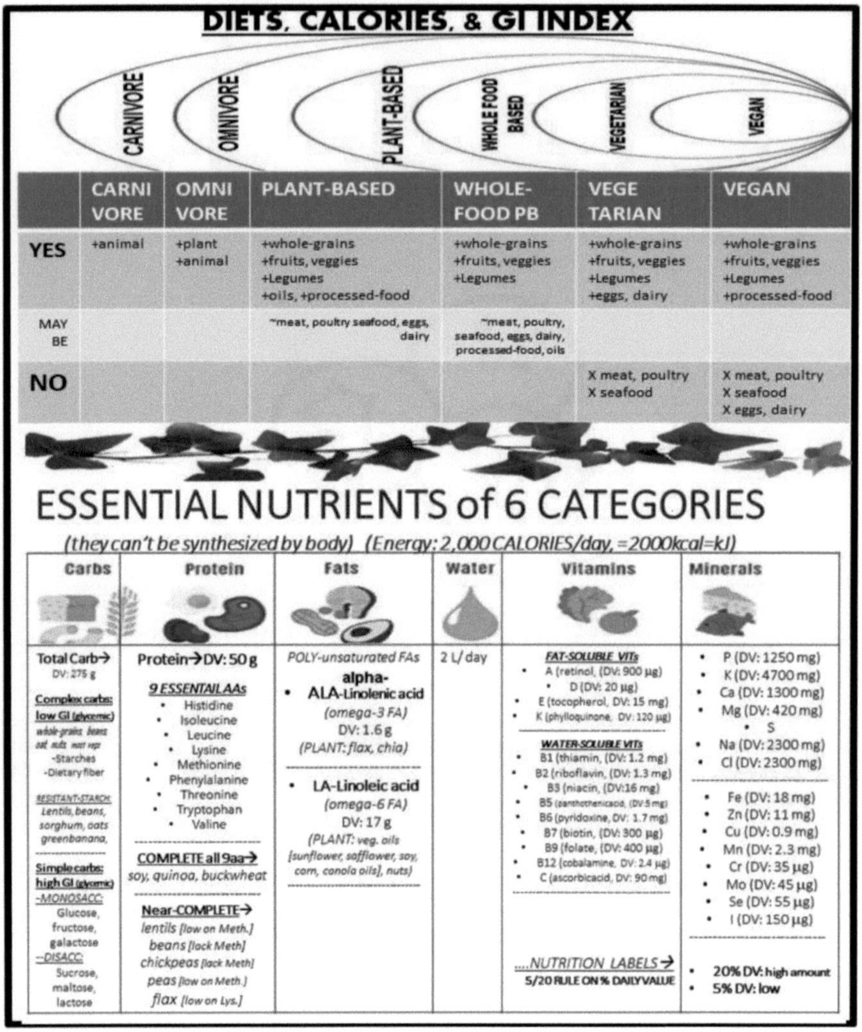

Fig. 11.1 Infographics of this chapter

11.2 Student Learning Outcomes (SLOs) of the Chapter

- Explain the benefits of plant-based diets.
- Identify essential nutrients of carbohydrates, proteins, fats, and minerals.
- Distinguish between blue zones and un-blue zones, their diets, and longevity.
- Identify how diet affects greatest longevity in humans.

11.3 Concepts of the Chapter

11.3.1 Nutrition Fact Labels

Food manufacturers are required to list all ingredients and **nutritional values** on foods by Food and Drug Administration (FDA) (Fig. 11.2). Furthermore, the content of sugar, fat, and salt levels is also included in the **food labels**. The percent daily values (% DV) for each nutrient are based on the recommended daily percentage on a standard adult's 2000-calorie diet. Food crop plants could provide a significant % DV for human diets.

The rule of thumb is that 5% DV is low and 20% DV is high in a specific nutrient. While a 100-calorie per serving is good, a 300-calorie per serving is considered high-calorie level (Fig. 11.2).

For more on this topic, see Altman (1998) and many references therein.

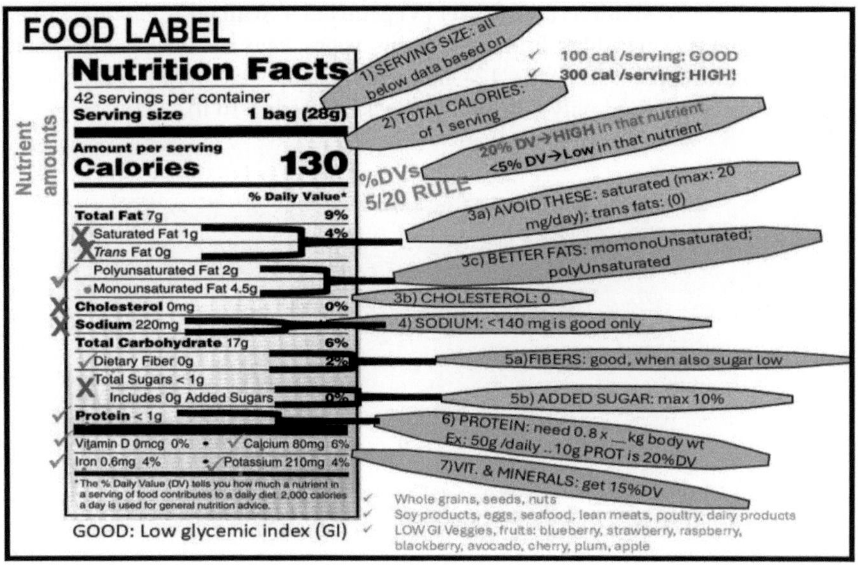

Fig. 11.2 Nutrition facts labels. Food labels start with serving size information and continue with total calories per serving, and all essential and beneficial nutrient information together with their percent daily values (% DV)

11.3.2 Plant-Based Food and Diets

Plant-based diets, vegan, vegetarian, and whole-food plant-based diet, center around plant foods such as vegetables, fruits, whole grains, nuts, seeds, and healthy plant oils (Figs. 11.1, 11.3).

Crop species such as soybeans, quinoa, and buckwheat have full complement of amino acids. While plants have B9 vitamin (folate), A, C, E, K, D vitamins, iron (non-heme, less avail), they may be missing vitamins B12, K2, and essential fatty acids (EPA and DHA). Plants are also rich in fiber, antioxidants, pectin, and potassium.

Plant-based diets are renowned for human health and environmental advantages. Some of the most followed plant-based diet varieties include:

Part 1: Vegan Diet: This diet includes whole grains *(barley, brown rice, oats, buckwheat, einkorn, teff, wild rice, red rice, triticale, quinoa, sorghum, spelt, bulgur),* veggies, fruits, and legumes but excludes all animal products (no meat, dairy, eggs, or seafood).

Part 2: Vegetarian Diet: This diet includes whole grains *(barley, brown rice, oats, buckwheat, einkorn, teff, wild rice, red rice, triticale, quinoa, sorghum, spelt, bulgur)*, veggies, fruits, legumes, eggs, and dairy but excludes most animal products (no meat, poultry, or seafood).

Part 3: Pescatarian Diet [Vegetarian + Seafood]: This diet includes whole grains *(barley, brown rice, oats, buckwheat, einkorn, teff, wild rice, red rice, triticale, quinoa, sorghum, spelt, bulgur)*, veggies, fruits, legumes, eggs, fish, and seafood but excludes dairy products, meat, and poultry.

Part 4. Pesca + Mediterranean Diet [Vegetarian + Seafood + Fasting 12 h+]: This diet includes vegetables, fruit, nuts, grains, beans, eggs and omega-3 fatty acid-rich seafood (salmon, sardines) + 12+ hours fasting (only eat during 8 h period so that body digesting food only 8 h, and no digesting during 16 h).

Fig. 11.3 Plant-based diets focus mainly on plant food sources, including vegetables, fruits, whole grains, nuts, seeds, and plant oils. Including plenty of colors is also part of plant-based diets. (Image: G. Hacisalihoglu, PhD)

Part 5: Whole-Food [Minimally Processed] Plant-Based Diet: This diet includes unprocessed plant-based foods such as whole grains *(barley, brown rice, oats, buckwheat, einkorn, teff, wild rice, red rice, triticale, quinoa, spelt, and bulgur)*, veggies, fruits, and legumes.

Part 6: Flexitarian Diet: This diet includes mainly plant-based foods and seafood but excludes meat and dairy products.

More people choose plant-based diets due to proven benefits, including lowering cholesterol, lowering caloric intake, and their environmental friendliness. Furthermore, plant-based milk is also a growing industry, including soy milk, almond milk, coconut milk, rice milk, oat milk, and flaxseed milk, among others.

For more on this topic, see Haytowitz et al. (2019) and many references therein.

11.3.3 Superfoods (Plants): Antioxidant Protectors of Cells

Superfoods, nutrient-rich food sources, are renowned for outstanding health benefits. Their antioxidant molecules eliminate harmful free radicals and protect cells. What are the benefits of plant-based superfoods? Some benefits include high nutrient content, high vitamin content, high antioxidant content, heart health, and disease prevention, among others.

Top plant-based superfoods are as follows: blueberries, raspberries (for antioxidants and fiber), kale, spinach (for iron, calcium, vitamins A, C, K), avocado (for healthy monounsaturated fats [MUFAs]), legumes such as beans and lentils (for plant protein [9 g protein per oz], fiber, and folate content) (Table 11.1; Fig. 11.4). Colorful plants are rich in colors such as red for anthocyanins (tomato, red onions), orange yellow for carotenoids (pumpkin, carrot, sweet potato), and purple blue for anthocyanins.

For more on this topic, see Grusak and DellaPenna (1999), FAO (2016, 2017), Gong et al. (2020), and many references therein.

11.3.4 Glycemic Index (GI): Food Glucose Impact (Low GI Doesn't Cause Sugar Spikes)

Glycemic index (GI): A classification system that ranks carb-rich foods on a scale of 0–100 according to their impact on blood sugar levels. Eleven major groups of plants with sugar-lowering attributes are as follows: cinnamon, cloves, rosemary, oregano, sage, curry leaves, garlic, ginger, turmeric, and cayenne (Fig. 11.5, Table 11.1).

For more on this topic, see Philippou (2016) and many references therein.

Table 11.1 The top plant-based superfoods packed with plant-based proteins, healthy fats (mono-unsaturated, poly-unsaturated), fibers, mineral nutrients, vitamins, antioxidants, and their health benefits

Plant-based superfoods	Main health benefits
Algae/seaweed (spirulina)	Rich in anticancer, weight loss, blood pressure lowering
Almonds	Lowers sugar levels, vit. E, P, K, Mg, Ca
Avocados	Rich in protein, healthy fats, fiber, K; support heart, brain, eye health
Beets	Rich in nitrates, relaxes blood vessels, improves blood flow and athletic performance
Blueberries	Rich in antioxidants, support brain, anti-inflammation, heart health, vit. C
Blackberries	Rich in protein, anthocyanins, heart health, bones, Mn, K
Broccoli	Rich in fiber, vit. C, K, anti-inflammation, lower cancer risks
Chia seeds	Rich in omega-3s, fiber, protein; supports digestion, heart
Flaxseed	Rich in omega-3s (polyunsat. fats): alpha-linolenic acid (ALA), cholesterol lowering, anticancer
Kale	Rich in vit. A, C, K, supports detox, bone, and eye health, anticancer
Kiwi	Rich in protein, vit. C, K, blood pressure control
Quinoa	Complete plant protein, rich in fiber, aids sugar control
Spinach	Rich in iron, folate, antioxidants, potassium, magnesium, immunity, eye
Sweet potatoes	Rich in B-carotene (vit. A), fiber, antioxidants; support eye, gut
Turmeric	Anti-inflammatory, antioxidant, supports joint, immune health
Walnuts/almonds	Anti-inflammatory, antioxidant, supports joint, immune health
Tomato	Rich in antioxidants, Fe, K, C, vit. A
Turmeric	Anti-inflammatory, antioxidant, supports joint, immune health
Walnuts	ALA omega-3s, heart, cholesterol lowering
Whole grains (oats, barley, brown rice, bulgur)	Complex carbs, heart healthy, lowers sugar levels

Fig. 11.4 Representative superfoods include blueberries (lower diabetes risks), blackberries (lower cancer risks), strawberries (lower cholesterol and blood pressure), and avocados (lower cholesterol and reduce heart disease). (Image: G. Hacisalihoglu, PhD)

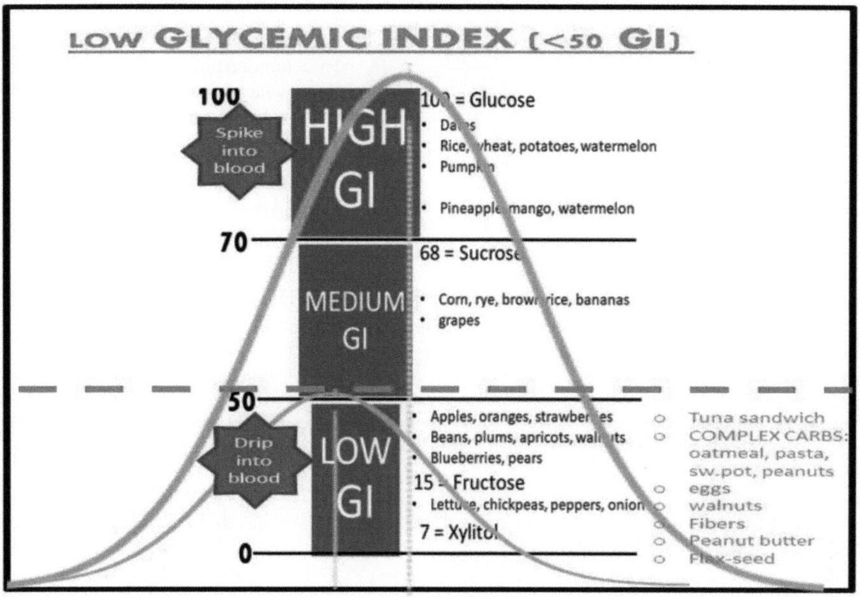

Fig. 11.5 Glycemic index (GI) is a classification that estimates how quickly food is absorbed into the blood. Low GI (<50) plants include berries, apples, carrots, broccoli, legumes, and whole grains such as quinoa and brown rice. See Hacisalihoglu (2021) for many references therein for further details

11.3.5 Support Greatest Longevity (Super Agers) and Antiaging Plants

Plants, **rich in polyphenols**, are renowned for promoting great human longevity and antiaging (Fig. 11.6). For example, blue zones mainly use whole-food plant-based diets and have long-lived centenarian populations.

Plant phytochemicals that promote healthy aging are polyphenols. Polyphenols contain properties of antioxidants to fight free radicals and restore the functions of mitochondria. Some examples of polyphenol-rich plant species are strawberries, hazelnuts, blueberries, olives, flaxseed, red onion, spices, herbs, grapes, soybeans, plums, almonds, and black beans, among others (Fig. 11.6).

Part 1. Reaching 100: Longevity is antiaging- and lifespan-promoting characteristics of centenarians who live 100+ years old of age. For example, Japan is known for having the most centenarians. Both genetic factors (longevity genes, epigenetic alterations, and telomeres) and environmental factors (diet, exercise, air quality, and water quality) may contribute to human longevity (Fig. 11.7). There are plant-based factors that contribute to human longevity due to plants' beneficial content of antioxidants, fiber, and phytochemicals (Fig. 11.7).

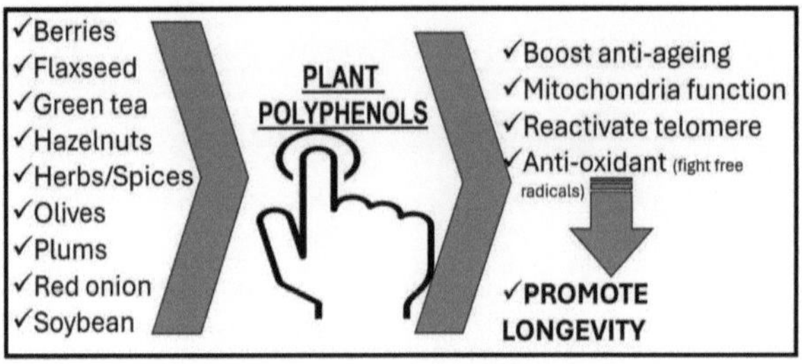

Fig. 11.6 Some of the representative longevity plant species include green tea, olives, berries, garlic, plums, ginger, and much more. These plants contain high polyphenols to boost antiaging and mitochondria function and potentially reactivate telomeres

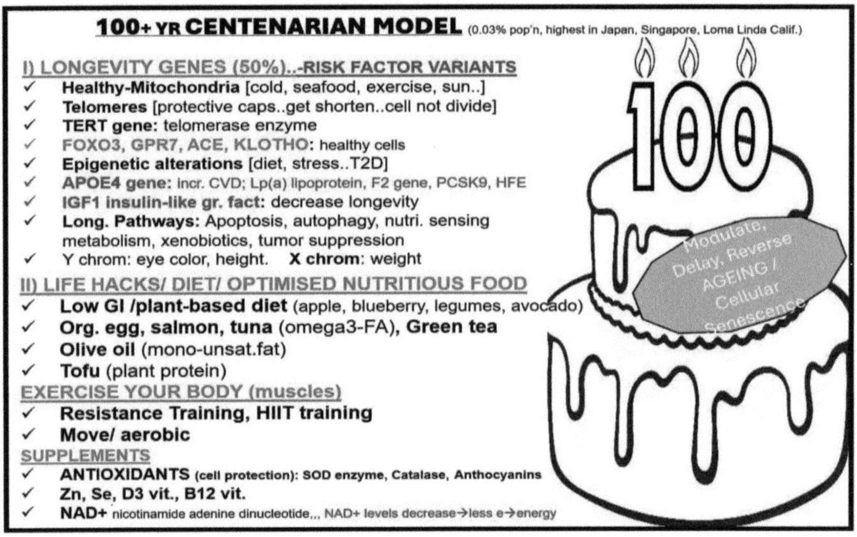

Fig. 11.7 Reaching 100. A 100-plus year centenarian model to achieve longevity, including genetics, diet, optimized nutrition, and exercise routines. Japan, Hong Kong, Singapore, South Korea, Australia, and New Zealand have the highest human life expectancies

While **plants have modular indeterminate continuous growth**, animals and humans have determinate unitary growth. Therefore, animals and humans stop growth at maturity, while plants can add repeated units (modules) continuously throughout life. As a consequence, modularity gives plants more resilience and longer life span like.

- **Oldest Living Plant:** A bristlecone pine (Methuselah) in California, USA (4850 years old). *Pinus longaeva* located in 3000+ m elevation in White Mountains.

- **Oldest Living Animal:** A giant tortoise (Jonathan) aged 193 (1832—). *Aldabrachelys gigantea* currently lives in Saint Helena islands between Brazil and Angola.
- **Oldest Living Humans (1 centenarian/ per net zer5,000 people):** M: NETO, Joao of Brazil age 113 (1912—) and F: CATERHAM, Ethel of the UK age 116 (1909—).

For more on this topic, see Buettner (2009), Buettner and Skemp (2016), FAO, Olshansky et al. (2001), Tilman and Clark (2014), and many references therein.

11.3.6 Type 2 Diabetes (T2D) and Plants

Glucagon-like peptide-1 (**GLP1**) hormone naturally helps control blood sugar and fullness. However, GLP1 secretion may be impaired due to T2D (Fig. 11.8). Whole grains with bran and embryo have the potential to lower heart disease and diabetes. GLP1 is released in response to food, then reduces appetite, and increases insulin levels.

Major groups of plant species known to **lower blood sugar levels** are as follows: Aloe vera, cinnamon, ginger, onions, fenugreek, ginseng, bitter melon, garlic, and turmeric. The T2D medicine metformin was originally synthesized from the plant species goat's rue (French lilac, *Galega officinalis*).

For more on this topic, see Willcox et al. (2008), McCouch et al. (2013), Balick and Cox (1996), Eichhorn et al. (2012), Reece et al. (2014), FAO (2017, 2024), FAOSTAT (2023), Hacisalihoglu (2011, 2021, 2022), Harshberger (1896), Judd et al. (2015), Prance et al. (2005), Ridolfi (2019), Schmidt and Cheng (2017), Simpson and Ogorzaly (2013), Haytowitz et al. (2019), USDA (2023), USDA-NRCS (2000), Weber et al. (2024), and many references therein.

For more on this topic, see Amos et al. (1997), FAO (2016), Gong et al. (2020), and many references therein.

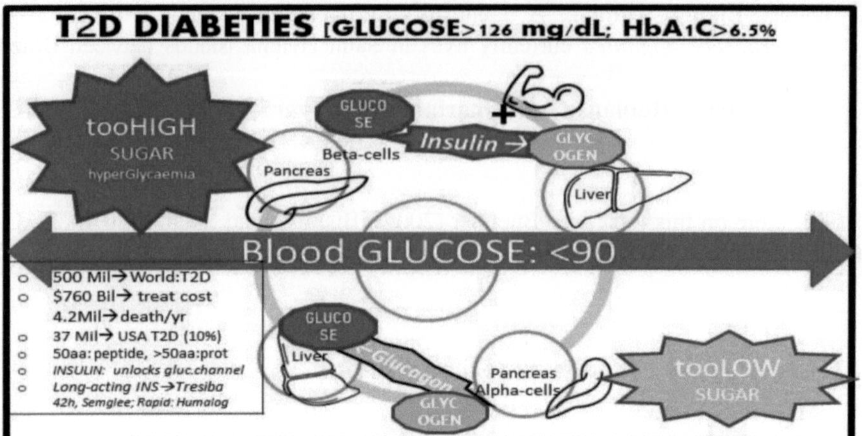

ORGAN	T2D RESULTS	CLASS/DRUG/EFFECT
PANCREAS -Beta-cell fail.	-low insulin prod	~DPP4-enzy-inhibitors ~Sulfonyl-Urea ...+++insulin prod.
LIVER	+high GLUCOSE prod.	~Biguanides: METFORMIN ..---lower glucose-prod., insul.resist.
KIDNEY	-kidney dis.	~SGLT2-inhibitors ~sodium-gluc trans [delay glucose absorb) ...+++Januvia: 1% lower H1b
GUT [DIGEST.TRACK]		~GLP1-receptor agonists ~glucagon-like peptide transporter +++Ozempic: slow carb.digest.
MUSCLE (&body fat)	-low uptake/utilization of GLUCOSE	+wt.lift +exercise (lower insul.resist)
HEART	+CVD risk +heart fail.	
BLOOD &eye/ foot	+GLUCOSE	+insulin inject.

Fig. 11.8 Type 2 diabetes (T2D) starts with insulin resistance, not effectively moving glucose out of blood to the other cells, and glucose accumulation in the blood. A number of plant species are renowned for their potential lowering blood glucose levels, including cinnamon, ginseng, and bitter gourd

11.4 Plant Profiles Bingo: How Many Can You Identify?

<table>
<tr><td>(P101) 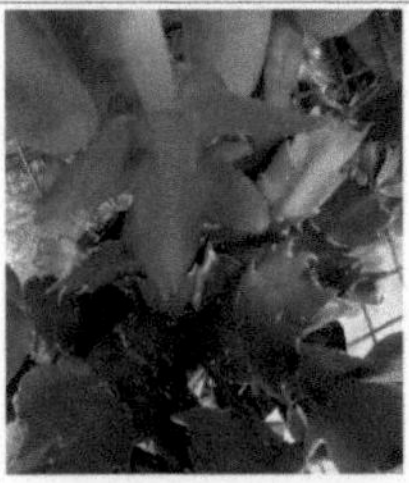
Schlumbergera bridgesii</td><td>(P102) 
Tephrocactus articulatus</td></tr>
<tr><td>(P103) 
Theobroma cacao</td><td>(P104) 
Amorphophallus titanum</td></tr>
<tr><td>(P105) 
Trifolium repens</td><td>(P106) 
Hydrangea macrophylla</td></tr>
<tr><td>(P107) 
Victoria cruziana</td><td>(P108) 
Viola wittrockiana</td></tr>
<tr><td>(P109) 
Vitis vinifera</td><td>(P110) 
Welwitschia miriabilis</td></tr>
</table>

11.5 Takeaway Questions

1. What is a plant-based diet?
2. What is the function of antioxidants in certain plant species such as berries?
3. Could a plant-based diet lower heart disease?
4. What makes chia seeds a plant-based superfood?

11.6 Now You Know the Answers

1. A plant-based diet mainly focuses on plant sources such as vegetables, fruits, and whole grains.
2. Antioxidants fight, neutralize free radicals, and therefore help reduce oxidative stress.
3. Certainly. Plant-based diets can reduce cholesterol levels and blood pressure levels, and therefore they can help with lowering heart disease.
4. Chia seeds are rich in proteins, fibers, and omega-3 fatty acids which support healthy immune systems.

References

Altman TA (1998) FDA and USDA nutrition labelling guide, *Checklist and Regulation*. Technomic Publishing Company, Inc, Lancaster

Amos A, McCarthy D, Zimmet P (1997) The rising global burden of diabetes and its complications: estimates and projections to the year 2010. Diabetic Med 14:S1–S85

Balick MJ, Cox PA (1996) People, plants, and culture: the science of ethnobotany, New York

Buettner D (2009) How to live 100+. TED Ideas Worth Spreading. https://www.ted.com/talks/dan_buettner_how_to_live_to_be_100

Buettner D, Skemp S (2016) Blue zones: lessons from the world's longest lived. Am J Lifestyle Med 10:318–321

Eichhorn SE et al (2012) Raven biology of plants, 7th edn. Macmillan

FAO (2016) International year of legumes. Nutritious seeds for a sustainable future, Rome. www.fao.org/pulses-2016

FAO (2017) The future of food and agriculture: trends and challenges. Food and Agriculture Organization of the United Nations, Rome

FAO (2024) Food and Agriculture Organization of the United Nations. Current market situation and medium-term outlook 1 Aug 2025. http://www.fao.org

FAOSTAT (2023) FAO statistical databases. Food and Agriculture Organization of the United Nations, Rome. https://www.fao.org/faostat

Gong X et al (2020) Effects of phytochemicals from plant-based functional foods on hyperlipidemia and their underpinning mechanisms. Trends Food Sci Technol 103:304–320

Grusak MA, DellaPenna D (1999) Improving the nutrient composition of plants to enhance human nutrition and health. Ann Rev Plant Physiol Plant Mol Biol 50:133–161

Hacisalihoglu G (2011) Plant biology concepts and laboratory manual. Kendall Hunt Publishing, Iowa. ISBN-13: 9780757589133. https://www.amazon.com/Plant-Biology-Concepts-Laboratory-Manual/dp/0757589138

Hacisalihoglu, G. (2021). From growing to biology: plants 1. Florida State University. 157. ISBN13: 9781736577912

Hacisalihoglu G (2022) Level up college success: research, writing, and more key skills for STEM learners. ISBN13: 9798218066260. www.amazon.com/author/drhaci

Harshberger JW (1896) The purposes of ethnobotany. Bot Gaz 21:146–154

Haytowitz DB et al (2019) USDA National Nutrient Database for standard reference, Legacy Release. Nutrient Data Laboratory, Beltsville Human Nutrition Research Center. https://data.nal.usda.gov/dataset/usda-national-nutrient-database-standard-reference-legacy-release

Judd WS et al (2015) Plant systematics: a phylogenetic approach, 4th edn. Sinauer Associates, Sunderland

McCouch S et al (2013) Feeding the future. Nature 499:23–24

Olshansky SJ, Carnes BA, Desesquelles A (2001) Demography: prospects for human longevity. Science 291:1491–1492

Philippou E (2016) The glycemic index: applications in practice. CRC Press, London

Prance GT et al (2005) The cultural history of plants. Routledge, New York

Reece JB et al (2014) Campbell biology, 10th edn. Pearson, Boston

Ridolfi R (2019) SDGs for people, planet and prosperity. www.openknowledge.fao.org

Schmidt BM, Cheng DMK (2017) Ethnobotany: a phytochemical perspective. Wiley

Simpson BB, Ogorzaly MC (2013) Plants in our world: economic botany, 4th edn. McGraw-Hill

Tilman D, Clark M (2014) Global diets link environmental sustainability and human health. Nature 515:518–522. https://doi.org/10.1038/nature13959

USDA (2023) National Agriculture Statistics Service. https://www.nass.usda.gov/

USDA-NRCS (2000) The PLANTS database. National Resources Conservation Service. https://plants.usda.gov/

Weber C, Simnitt S, Wakefield H, Wechsler S, Swearingen B (2024) Fruit and tree nuts outlook: March 2024. United States Department of Agriculture, Washington, DC. https://www.ers.usda.gov/publications/pub-details?pubid=108880

Willcox BJ, Donlon TA, He Q, Chen R, Grove JS, Yano K et al (2008) FOXO3A genotype is strongly associated with human longevity. Proc Natl Acad Sci USA 105:13987–13992

Chapter 12
Future Plants Now: Feeding the Twenty-First Century

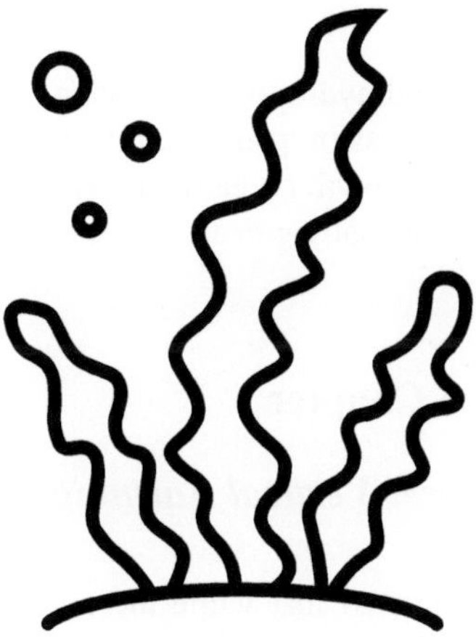

Future Plants Now: Feeding the Twenty-First Century

Contents

G. Hacisalihoglu, *Super Greens*, https://doi.org/10.1007/978-3-032-17047-7_12

12.1 Big Picture: Quick Read Chapter Infographic

12.2 Student Learning Outcomes (SLOs) of the Chapter

- Explain vertical farming and hydroponics growing.
- Identify effect of climate stress on plant growth and nutrition.
- Distinguish between plant growth on earth and space in zero gravity.
- Identify trendy microgreens, baby greens, and sprouts plants.

12.3 Concepts of the Chapter

12.3.1 Grow Anywhere: Vertical Farming and Hydroponics

By the **year 2050**, it is estimated that while there will be less arable land (from 0.20 ha per capita to 0.15 ha per capita), there will be risen world population (10 billion), global water demand, increased loss to pests (over 20%), and increased average global temperatures (from 14.7 C to 16 C).

Hydroponics, a soilless plant growing technique, can be utilized in vertical or horizontal growing spaces (Figs. 12.2 and 12.3). Nutrients are provided to plants via water solutions (Fig. 12.2). Some of the advantages of hydroponics growing are optimal conditions, faster growth, less space, extended seasons, and any location. Some of the substrates used in hydroponics are peat moss and perlite mix. Among the nutrients, P helps with bloom, while N and K help with vegetative growth of crop plants. The most popular hydroponically grown plants are red kale, basil, dill, lettuce, strawberries, oregano, peppers, tomatoes, and microgreens. Some of the hydroponics systems include floating raft system, nutrient film system, and fine mist aeroponics, among others.

For more on this topic, see UN (2020), Alexandratos and Bruinsma (2012), Ferl et al. (2002), Bouis and Welch (2010), Benke and Tomkins (2017), Beacham et al. (2019), Blair (2013), Bar-On et al. (2018), Avnee et al. (2023), Balick and Cox (1996), Eichhorn et al. (2012), Reece et al. (2014), FAO (2017, 2024), FAOSTAT

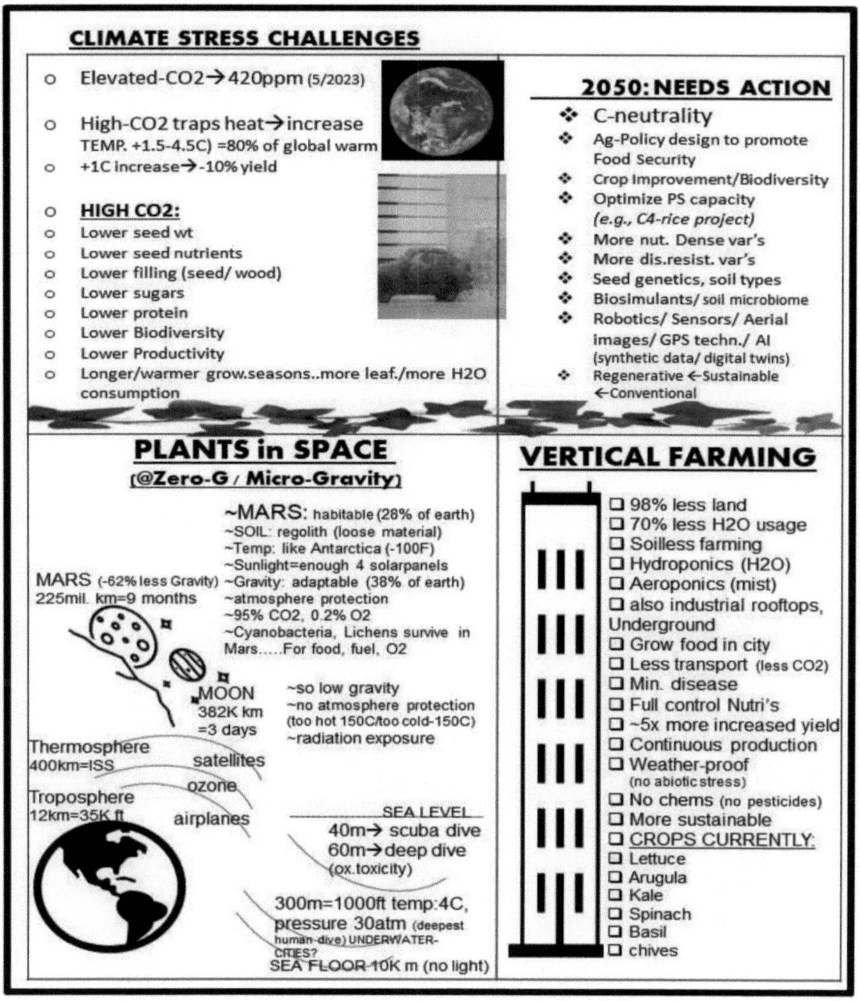

Fig. 12.1 Infographics of this chapter

(2023), Hacisalihoglu (2011, 2021, 2022), Harshberger (1896), Judd et al. (2015), and many references therein.

In **vertical farming** (aka urban farming, indoor farming) is a type of hydroponics using buildings or shipping containers to grow plants with most optimization. Blue and red LED lights help speed up indoor plant growth by helping with germination, flowering, chlorophyll synthesis, and photosynthesis (Fig. 12.3a). Some of the advantages of vertical farming are less land use, less water use, minimal labor use, no pesticides, better flavor, farm-fresh, and 24/7 growing with maximum crop yield.

For more on this topic, see Despommier (2010) and many references therein.

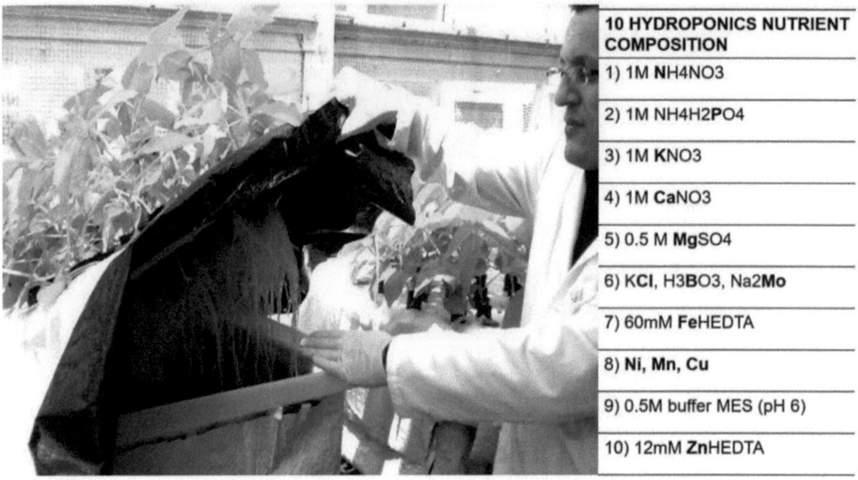

Fig. 12.2 Hydroponics solutions of water-based nutrient mixtures. See Hacisalihoglu and Kochian (2003) for many references therein for further details. (Image: G. Hacisalihoglu, PhD)

Fig. 12.3 (**a**) Vertical farming is one of the newest techniques to grow plants vertically indoors; (**b**) hydroponically grown lettuce in a soilless nutrient solution environment. (Image: G. Hacisalihoglu, PhD)

Part 1: Trendy Greens Ingredients (Sprouts and Microgreens): It has been popular recently to grow trendy greens, including sprouts, microgreens, and baby greens for rich nutrient content and flavor purposes (Fig. 12.4). These are younger versions of mature crops and vegetables which are harvested prematurely for two to 28 days. The most common plant species used for microgreens are arugula, kale, radish, cabbage, and peas.

SEEDS	SPROUTS	MICROGREENS	BABY GREENS	MATURE PLANTS
	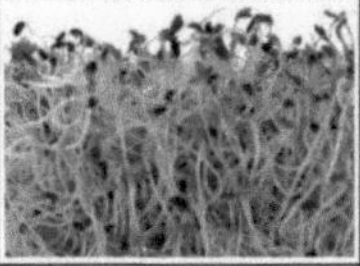			
HARVEST→	Germinated seeds ~5 cm	Fully developed cotyledon leaves + 1st leaves ~10 cm	Unmature leaves of vegetables/ herbs ~15 cm	Mature ~>20 cm-
0 d	Ready: 2-7 d Seed leaves +Rootlets	Ready: 7-14 d Shoots + Cotyledons	Ready: 21-28 d Leaves	Ready: >30-60 d Leaves
BENEFITS →	Nutrient-dense (Ca, Fe, Mg, Zn) & Antioxidants *-Alfalfa sprouts, Mung bean sprouts*	Flavor, delicate, crisp flavor, Nutrient and Vitamins rich *-Arugula, Red Kale, Radish, Red Acre Cabbage, Pea*	For smoothies, salads High in nutrients *-Arugula, Kale, Lettuce, Spinach, Collard greens*	*Leafy greens: Arugula, Basil, Kale, Lettuce, Spinach, Strawberries, Peppers, Bok choy*

Fig. 12.4 Representative trendy greens ingredients are sprouts, microgreens, and baby greens. While sprouts involve whole plants, microgreens and baby greens involve only shoot parts of the plants

For more on this topic, see Tilman and Clark (2014), Tilman et al. (2011), Spencer et al. (2017), Schroeder et al. (2013), Saltzman et al. (2017), Ray et al. (2013), McCouch et al. (2013), Massawe et al. (2016), Hacisalihoglu (2021), and many references therein.

12.3.2 Climate Stress: Increasing CO_2 Levels Decrease Plant Ionome Levels

It is estimated that global food crops production levels may decline in upcoming decades due to climate stress, heat, drought, floodings, and potential water supplies issues (Fig. 12.5). Moreover, earth's population is expected to reach **10 billion by 2050,** which is another **30% more than the current year of 2025**. Extreme drought and heat can lower yield and nutrients of plant species.

12.3.3 Carbon Cycle and Net-Zero Carbon Neutral

Carbon is found in all living things, including plants. Carbon cycles between atmosphere and earth. Green plants move CO_2 from atmosphere to biosphere and

Fig. 12.5 Saguaro cactus can tolerate extreme levels of heat and drought stress via storing water in their stem. Saguaros protected from water loss with a waxy cuticle and spines. (Image: G. Hacisalihoglu, PhD)

use it in the most important reaction in biosphere to make food and carbohydrates in photosynthesis (Fig. 12.6).

Part 1. Net-Zero Future: There is significant progress on reaching **net zero** and reducing greenhouse gases. Carbon neutrality is achieving the net-zero carbon footprint between released and offset carbon levels (Fig. 12.6). Net zero could also be achieved via carbon removal sequestration from source to various carbon sinks. Net zero is a sustainable long-term viable solution.

Part 2: Carbon Source to Carbon Sink

Carbon Source(s): Volcanoes, respiration, burning fossil fuels, manufacturing industry, construction processes, cement production, forest clearing, and electricity production are common carbon sources.

Carbon Sink(s): Oceans (CO_2 dissolving), photosynthesis, forests, soils, rocks, and artificial sequestration of carbon capture are common carbon sinks.

For more on this topic, see Garcia-Casal et al. (2017), Godfray et al. (2010), Hickey et al. (2019), Mapstone et al. (2022), Tilman and Clark (2014), Tilman et al. (2011), Spencer et al. (2017), Schroeder et al. (2013), Saltzman et al. (2017), Ray et al. (2013), McCouch et al. (2013), Massawe et al. (2016), Hacisalihoglu (2021), McCouch et al. (2013), Ray et al. (2013), and many references therein.

12.3.4 Bioenergy Plants: Bioethanol and Biodiesel (Fossil Fuel Alternatives)

Biofuels, alternative fuel sources to fossil fuels, are renewable sources of energy derived from plant biomass or algae. There are two common bioenergy sources, including bioethanol and biodiesel (Fig. 12.7). For more on this topic, see Kole et al. (2012) and many references therein.

Fig. 12.6 Overview of the carbon cycle. Carbon, chemical backbone of life, makes up the food and provides energy. Carbon is released to the atmosphere via fossil fuel burning

One issue with bioenergy plants is the possibility of food crops. Subsequently, bioenergy production could be in competition with human foods. Alternatively, cover crops could be used in biofuels, including clover, winter legumes, or algae alternatives.

For more on this topic, see Adeniyi et al. (2018), Mapstone et al. (2022), Ainsworth and Long (2021), Hacisalihoglu (2021), and many references therein.

Biofuel: bioethanol (for light vehicles to reduce fossil fuel reliance)	Biofuel: biodiesel (for heavy vehicles to reduce fossil fuel reliance)
➢ "Renewable" fuel from biomass to create energy: fermentation of sugar crops to produce bioethanol ➢ **FEEDSTOCKS USED:** ➢ Corn ➢ Grass ➢ Sugar cane ➢ Micro-Algae ➢ CONCERN: impact on food prizes. Less energy as a fuel. E85 (85% ethanol) cost: $2.31/ gallon E10 (10% ethanol) cost: $2.81/ gallon ➢ 2025 Trends: environmentally friendly: Carbon neutral	❖ "Renewable" fuel from biomass to create energy: Transesterification reactions of oils to produce biodiesel ❖ **FEEDSTOCKS USED:** ❖ Cooking oils ❖ Palm oil ❖ Coconut ❖ Castor ❖ Soybean ❖ Micro-Algae (GMO) for higher yields ❖ CONCERN: high production costs for B100 (100% biodiesel, $5/ gallon) ❖ 2025 Trends: environmentally friendly: Carbon negative

Fig. 12.7 Biofuels create fuel energy from biomass of feedstocks. Biofuels are renewable energy sources. Common plant species used for biofuels are corn, soybeans, and sugarcane. Biodiesel is typically made from vegetable oils

Fig. 12.8 Duckweed (*Lemna minor*) has the potential for future food and fuel source due to its fast growth, bioactive metabolite compounds, and plant proteins. Duckweed is the smallest flowering plant and carries the smallest flower

12.3.5 Duckweeds: Future Food, Biofuel, and Fastest-Growing Flowering Plant

Duckweed (*Lemna minor*) is very small aquatic and fast-growing flowering plant species in freshwater ponds and lakes (Fig. 12.8). Its large volume and biomass make duckweed a great potential candidate for biofuels. Duckweed is edible and has rich protein content. Therefore, duckweed can be further explored for food, fuel, and oil production purposes. For more on this topic, see Pagliuso et al. (2022) and many references therein.

12.3.6 Cyanobacteria and Lichens

Both cyanobacteria and lichens are resilient to environmental stress. There is a cyanobacteria strain, namely spirulina, used as food source. Cyanobacteria can both carry out photosynthesis and fix atmospheric nitrogen (Fig. 12.9).

Part 1. Cyanobacteria (Many Classes): Spirulina (*Arthrospira* spp., *Limnospira* spp.) is edible for humans. Spirulina is also used in dietary supplements as a high protein food. For more on this topic, see Sarma 2012 and many references therein.

Part 2. Lichens: Lichen is an algae plus fungi composite organism. Lichens could be used for food sources, medicine, and air quality indicator organism. Lichens could clean air via absorption of air pollutants. For more on this topic, see Brodo et al. (2001) and many references therein.

12.3.7 Biofortification (BF)

Biofortification, process of enhancing nutritional content of food crops via agronomy, plant breeding, or genetics, aims to address global nutrient deficiencies for humans. Moreover, biofortification also has a target of enhancing nutrient content of whole crop plants more sustainably. The major nutrients targeted by biofortification include zinc (Zn), iron (Fe), and vitamin A, among others.

Fig. 12.9 Cyanobacteria has existed over 3.5 billion years and produced oxygen

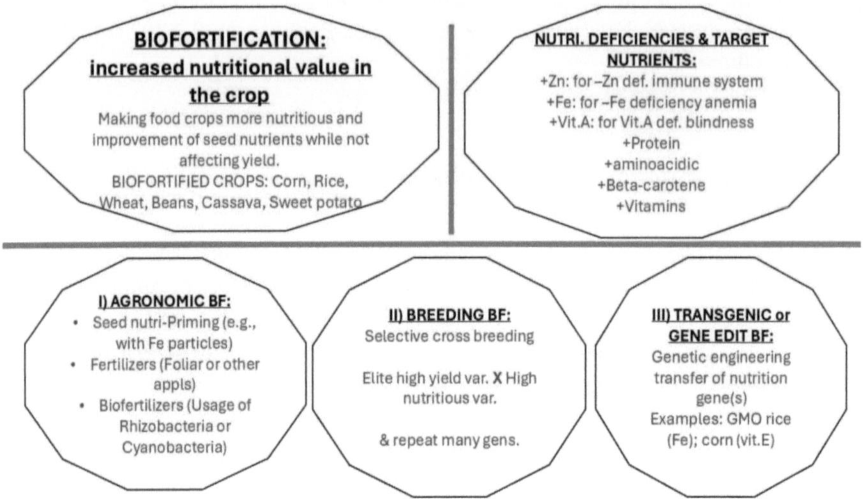

Fig. 12.10 Three major biofortification methods. BF: biofortification. Biofortification can improve micronutrient content of food crop plants

For more on this topic, see Singh et al. (2016) and many references therein.

Some of the major biofortification projects include golden rice with improved beta-carotene (vitamin A), wheat with enhanced Zn levels, beans with enhanced Fe levels, and sweet potato with enhanced vitamin A levels (Fig. 12.10).

12.3.8 Landscape Plants: Take It Outside

The four major groups of plant classifications are as follows:

Part 1: Garden Design and Master Plan: Perennials have distinct advantages, including deep roots, resilience, and presence without annual planting requirement. Common perennial flowering plants include yarrow, viola, hibiscus, daisy, and bee balm, among others.

- **Annual Life Span Plants (1-year)**: Wheat, rice, corn, petunias, marigolds
- **Biennial Life Span Plants (2-years)**: Onion, leek, parsley, cabbage, celery
- **Perennial Life Span Plants (>2 years)**: Artichokes, asparagus, chives, chrysantemum, hosta, peony, hydrangea, thrift, daylily, iris, lavender, bee balm, dianthus, palms, maple

For common landscape plant choices in Florida (Tallahassee, FL), growers could consider a wide variety of species, including hibiscus, crepe myrtle, azalea, lantana, and lily of the Nile (Table 12.1). For more on this topic, see Walton and Schiller (2007), Flint (1997), and many references therein.

Table 12.1 Common landscape plants to add to Florida gardens

Type	Name	Sci. name	Color	Characteristics
Shrub	Hydrangea	*Hydrangea macrophylla*	Blue, pink, white	Moist soil
Shrub	Crape myrtles	*Lagerstroemia indica*	Pink, purple, white	Vase shape
Shrub	Saw palmetto	*Serenoa repens*	Green foliage	Slow growing
Shrub	Coontie	*Zamia pumila*	Green foliage	Cycad, Florida native
Shrub	Lavender	*Lavandula angustifolia*	Purple	Fragrant, oil, bees
Shrub	Beautyberry	*Callicarpa americana*	Pink flowers	–
Tree	Apple tree	*Malus domestica*	Red, yellow fruit	Edible fruit, pruning
Tree	Blueberry	*Vaccinium corymbosum*	Blue fruit	Acidic soil
Tree	Magnolia	*Magnolia grandiflora*	White, pink, purple	Fragrant
Annual	Zinnia	*Zinnia elegans*	Red, pink, yellow, white	–
Annual	Marigold	*Tagetes patula*	Yellow, orange	Pest-repellent
Annual	Geranium	*Pelargonium* spp.	Red, pink, white	Scented leaves
Annual	Pansy	*Viola* × *wittrockiana*	White, blue, red, bicolor	
Annual	Petunia	*Petunia* spp.	Pink, purple, red, white	
Annual	Snapdragon	*Antirrhinum majus*	Pink, white, yellow, red	Flower snaps
Annual	Coleus	*Coleus scutellarioides*	Green, red, variegated	Foliage/leaves
Annual	Impatients	*Impatiens walleriana*	Pink, red, purple, white	Shade

Part 2: Lawn, Turf, and Natural Grass: (Table 12.2)

a. Lawn: Closely mowed (manicured) natural grasses with watering and fertilizing

b. Turf: Natural or artificial/synthetic turf for athletic fields

For more on this topic, see Christians et al. (2016) and many references therein.

Part 2: Bee-Friendly or Bird-Friendly Garden Plants: Bee pollinator plant species include lavender, marigold, salvia, aster, basil, and dahlia among others (Table 12.3).

For more on this topic, see Frey and LeBuhn (2015) and many references therein.

Part 4: Cut Flowers and Floriculture Crops: Cut flower production is over $400 million in the USA (Fig. 12.11). Some important factors in cut flowers production are day length, season, and greenhouse growth of annual and perennial species.

Table 12.2 Most popular turfgrasses, performance, and cultivars

	Turfgrass	Sci. name	Performance	Cultivar
In FLORIDA	**St. Augustine grass**	*Stenotaphrum secundatum*	Fast growth; drought tol.; low traffic	~'Palmetto', 'CitraBlue'
(warm climate)	**Bahiagrass**	*Paspalum notatum*	Weekly mowing; low water req.	~'Pensacola'
	Zoysiagrass	*Zoysia matrella*	Low maintanence; drought tol.	~'Empire', 'Eltoro'
	Bermudagrass	*Cynodon dactylon*	Golf courses; deep roots; high traffic	~'Latitude', 'Riviera'
	Centipedegrass	*Eremochloa ophiuroides*	Low mow req.	~'Centennial', 'TifBlair'
In theUSA	**Kentucky bluegrass**	*Poa pratensis*	Soft; high traffic; cool season	~'Midnight', 'Prosperity'
(Every climate), north	**Tall fescue**	*Festuca arundinacea*	Heat tolerant	~'Amity'
	Perennial ryegrass	*Lolium perenne*	Soccer fields; quick growth	~'Fireball' 'Champion'
	Buffalograss	*Bouteloua dactyloides*	Low water req., very high drought tolerant	~'Legacy', 'Turffalo'
In EU	**Bermudagrass**	*Cynodon dactylon*	Sports fields; sun req.	~'Tifway419', 'SR9554'
(Warm climate)	**Zoysia japanese**	*Zoysia japonica*	Shaded lawns; slow growth	~'Zenith'

Top **floriculture** crops include roses, tulips, orchids, lilies, mums, dahlias, zinnias, daisies, sunflowers, ranunculus, celosia, cosmos, peonies, and snapdragons. Cut flowers are used in floral designs in combination with stems and leaves.

For more on this topic, see Faust and Dole (2021) and many references therein.

12.3.9 Smart Farming: AI, IoT, Sensors, Drones, and Robotics for Real-Time Crop Monitoring

Smart farming integrates a combination of future technologies, including sensors, Internet of Things (IoT), drones, robotics, machine learning, data analytics, aerial imaging, and AI for continuous real-time crop monitoring (Fig. 12.12). Smart technologies help gather data on crop health, nutrition, climate, and soils for optimal crop growth. Other areas also of interest are yield prediction and resource optimization for crop growth.

For more on this topic, see Ferl et al. (2002), Bouis and Welch (2010), Benke and Tomkins (2017), Beacham et al. (2019), Blair (2013), Bar-On et al. (2018),; Avnee

Table 12.3 Common bee-friendly or bird-friendly garden plants. Over 150 crop species benefit from bee pollination

Bee-friendly	Sci. name	Bird-friendly	Sci. name
(White, yellow, blue) + honeys nectar		**(Red, orange)**	
Anemone bush	*Carpentira californica*	Agave	*Agave* spp.
Blackberry	*Rubus fruticosus*	Blueberry	*Vaccinium*
Bee balm	*Monardo* spp.	Cherry	*Prunus* spp.
Blueberry	*Vaccinium corymbosum*	Dogwood	*Cornus* spp.
Catnip	*Nepeta* spp.	Grape	*Vitis* spp.
Clover red, white	*Trifolium repens*	Hawthorn	*Crataegus* spp.
Coneflower	*Echinacea purpurea*	Hibiscus	*Hibiscus* spp.
Cosmos	*Cosmos bipinnatus*	Plum	*Prunus* spp.
Heather	*Calluna vulgaris*	Raspberries	*Rubus* spp.
Lavender	*Lavandula angustifolia*	Rose	*Rosa* spp.
Linden (ihlamur)	*Tillia cordata*	Thistle globe	*Echinops* spp.
Marigold	*Calendula officinalis*	Tulip tree	*Liriodendron tulipifera*
Orange	*Citus sinensis*	Willow	*Salix* spp.
Oregano	*Origanum vulgare*	Yew	*Taxus* spp.
Rosemary	*Rosmarinus officinalis*	Zinnia	*Zinnia* spp.
Sage	*Salvia* spp.		
Skyflower	*Duranta erecta*		
Sunflower	*Helianthus* spp.		
Thyme	*Thymus* spp.		
Wildflowers	*Various*		
Rhododendron [komar]	*Toxic "mad" honey*		

Fig. 12.11 Cut flowers commonly include rose varieties. (Image: G. Hacisalihoglu, PhD)

Fig. 12.12 Smart farming: AI, IoT, sensors, drones, and robotics for real-time crop monitoring, yield prediction, and resource optimization

et al. (2023), Balick and Cox (1996), Eichhorn et al. (2012), Reece et al. (2014), FAO (2017, 2024), FAOSTAT (2023), Hacisalihoglu (2011, 2021, 2022), Harshberger (1896), Judd et al. (2015), Prance et al. (2005), Ridolfi (2019), Schmidt and Cheng (2017), Simpson and Ogorzaly (2013), USDA (2023), USDA-NRCS (2000), Weber et al. (2024), and many references therein.

12.4 Plant Profiles Bingo: How Many Can You Identify?

(P111) 

Yucca carigida

(P112) 

Zamia pumila

(P113) 

Zantedeschia aethiopica

(P114) 

Zinnia elegans

(P115) 

Salvia rosmarinus

(P116) 

Matricaria chamomile

(P117) 

Plectranthus sp.

(P118) 

Opuntia littoralis

(P119) 

Zebrina pendula

(P120) 

Impatients walleriana

12.5 Takeaway Questions

1. How smart farming technologies improve food crop production?
2. Which turfgrass is suitable for Florida gardens?
3. Give an example of biofortification of food crop plants.
4. How to achieve "net-zero carbon"?

12.6 Now You Know the Answers

1. Smart farming technologies can help with continuous monitoring of crops, predicting diseases, and productivity.
2. Two turfgrass species popularly used in Florida are St. Augustine grass and Zoysia grass.
3. Golden rice is a genetically modified rice with high vitamin A content.
4. Net zero is about no increase in CO_2 levels and can be achieved by using more renewable energy sources, including solar, wind, and biofuels.

References

Adeniyi OM et al (2018) Algae biofuel: current status and future applications. Renew Sustain Energy Rev 90:316–335

Ainsworth EA, Long SP (2021) 30 years of free-air carbon dioxide enrichment (FACE): what have we learned about future crop productivity and its potential for adaptation? Glob Chang Biol. https://doi.org/10.1111/gcb.15375

Alexandratos N, Bruinsma J (2012) World agriculture towards 2030/2050, ESA working paper no.12-03. Food and Agriculture Organization of the United Nations

Avnee S et al (2023) Biofortification: an approach to eradicate micronutrient deficiency. Front Nutr 10:1233070

Balick MJ, Cox PA (1996) People, plants, and culture: the science of ethnobotany, New York

Bar-On YM et al (2018) The biomass distribution on earth. Proc Natl Acad Sci U S A 115:6506–6511

Beacham AM et al (2019) Vertical farming: a summary of approaches to growing skywards. J Hortic Sci Biotechnol 94:277–283

Benke K, Tomkins B (2017) Future food-production systems: vertical farming and controlled-environment agriculture. Sustainability: Sci Prac Policy 13:13–26

Blair MW (2013) Mineral biofortification strategies for food staples: the example of common bean. J Agric Food Chem 61:8287–8294

Bouis HE, Welch RM (2010) Biofortification—a sustainable agricultural strategy for reducing micronutrient malnutrition in the global south. Crop Sci 50:S20–S32

Brodo IM, Sharnoff SD, Sharnoff S (2001) Lichens of North America. Yale University Press, New Haven

Christians NE, Patton AJ, Law QD (2016) Fundamentals of turfgrass management. Wiley, Hoboken

Despommier D (2010) The vertical farm: feeding the world in the 21st century. Thomas Dunne Books, New York

Eichhorn SE et al (2012) Raven biology of plants, 7th edn. Macmillan

FAO (2017) The future of food and agriculture: trends and challenges. Food and Agriculture Organization of the United Nations, Rome

FAO (2024) Food and Agriculture Organization of the United Nations. Current market situation and medium-term outlook 1 Aug 2025. http://www.fao.org

FAOSTAT (2023). FAO statistical databases; Food and Agriculture Organization of the United Nations: Rome, Italy; https://www.fao.org/faostat

Faust JE, Dole JM (2021) Cut flowers and Foliages. CAB International, Oxfordshire

Ferl R et al (2002) Plants in space. Curr Opin Plant Biol 5:258–263

Flint HL (1997) Landscape plants for eastern North America: exclusive of Florida and the immediate Gulf Coast. Wiley, New York

Frey K, LeBuhn G (2015) The bee–friendly garden. Penguin Random House, New York

Garcia-Casal MN et al (2017) Staple crops biofortified with increased vitamins and minerals: considerations for a public health strategy. Ann N Y Acad Sci 1390:3–13

Godfray HCJ et al (2010) Food security: the challenge of feeding 9 billion people. Science 327: 812–818

Hacisalihoglu G (2011) Plant biology concepts and laboratory manual. Kendall Hunt Publishing, Iowa. ISBN-13: 9780757589133. https://www.amazon.com/Plant-Biology-Concepts-Laboratory-Manual/dp/0757589138

Hacisalihoglu, G. (2021). From growing to biology: plants 1e. Publisher: Florida State University. 157 pages. ISBN13: 9781736577912

Hacisalihoglu G (2022) Level up college success: research, writing, and more key skills for STEM learners. ISBN13: 9798218066260. www.amazon.com/author/drhaci

Hacisalihoglu G, Kochian LV (2003) How do some plants tolerate low levels of soil zinc? Mechanisms of zinc efficiency in crop plants. New Phytol 159:341–350

Harshberger JW (1896) The purposes of ethnobotany. Bot Gaz 21:146–154

Hickey LT et al (2019) Breeding crops to feed 10 billion. Nat Biotechnol 37:744–754

Judd, W. S. et al. (2015). Plant systematics: a phylogenetic approach. 4e. Sunderland: Sinauer Associates

Kole C et al (2012) Handbook of bioenergy crop plants. CRC Press, Boca Raton

Mapstone LJ et al (2022) Cyanobacteria and microalgae in supporting human habitation on Mars. Biotechnol Adv 59:107946

Massawe F et al (2016) Crop diversity: an unexploited treasure trove for food security. Trends Plant Sci 21:365–368

McCouch S et al (2013) Feeding the future. Nature 499:23–24

Pagliuso D et al (2022) Duckweeds as promising food feedstocks globally. Agronomy 12:796

Prance GT et al (2005) The cultural history of plants. Routledge, New York

Ray DK et al (2013) Yield trends are insufficient to double global crop production by 2050. Plos One 8:e66428

Reece JB et al (2014) Campbell Biology 10e. Pearson, Boston

Ridolfi R (2019) SDGs for people, planet and prosperity. www.openknowledge.fao.org

Saltzman A et al (2017) Availability, production, and consumption of crops biofortified by plant breeding: current evidence and future potential. Ann N Y Acad Sci 1390:104–114

Sarma TA (2012) Handbook of cyanobacteria. CRC Press, New York

Schmidt BM, Cheng DMK (2017) Ethnobotany: a phytochemical perspective. Wiley

Schroeder JI et al (2013) Using membrane transporters to improve crops for sustainable food production. Nature 497:60–66

Simpson BB, Ogorzaly MC (2013) Plants in our world: economic botany, 4th edn. McGraw-Hill

Singh U et al (2016) Biofortification of food crops. Springer, New Delhi

Spencer L et al (2017) Space pomology: dwarf plums for fresh food production. NASA Human Research Program Investigators' Workshop 257:0–3

Tilman D, Clark M (2014) Global diets link environmental sustainability and human health. Nature 515:518–522. https://doi.org/10.1038/nature13959

Tilman D et al (2011) Global food demand and the sustainable intensification of agriculture. Proc Natl Acad Sci USA 108:20260–20264

UN (2020) The sustainable development goals report. United Nations Department of Economic and Social Affairs (UN DESA)

USDA (2023) National Agriculture Statistics Service. https://www.nass.usda.gov/

USDA-NRCS (2000) The PLANTS database. National Resources Conservation Service. https://plants.usda.gov/

Walton D, Schiller L (2007) Natural Florida landscaping. Pineapple Press Inc, Sarasota

Weber C, Simnitt S, Wakefield H, Wechsler S, Swearingen B (2024) Fruit and tree nuts outlook: March 2024. United States Department of Agriculture, Washington, DC. https://www.ers.usda.gov/publications/pub-details?pubid=108880

Chapter 13
Case Studies in Plants to Implement in Education

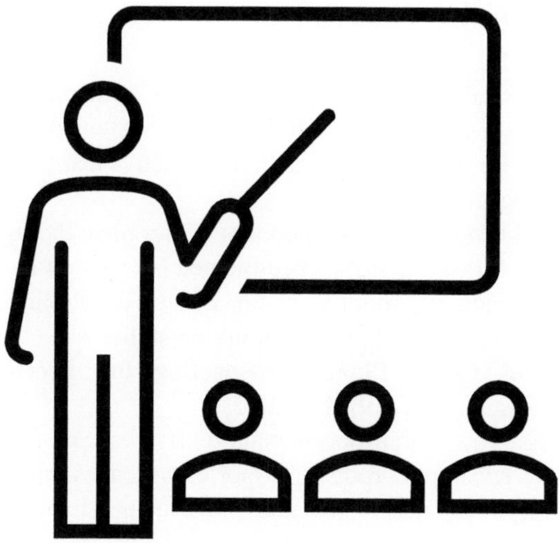

Case Studies in Plants to Implement in Education

Contents

G. Hacisalihoglu, *Super Greens*, https://doi.org/10.1007/978-3-032-17047-7_13

13.1 Big Picture: Quick Read Chapter Infographic

13.2 Student Learning Outcomes (SLOs) of the Chapter

- Understand the case study approach as a learning tool.
- Examine case studies in plant biology to identify and learn about key topics.
- Develop critical thinking skills in biology.
- Apply plant biology concepts to practice via analyzing cases (Fig. 13.1).

13.3 Concepts of the Chapter

Just what is a **case study** anyway? Case studies (CS) method involves real-life scenarios and encourages problem-based learning as well as decision-making. Furthermore, CS could help students' critical thinking, active learning, and appreciation for plant biology concepts.

How to get started? Here are some important CS for plant biology. We encourage learners to read the following plant science-themed CS by visiting the website links so that they could develop important plant concepts by developing mastery learning in depth and connections via higher-order thinking skills.

CS 1: Fields of Gold: Plants Prospecting for Precious Metals (by Hacisalihoglu and Strickland 2019) (Fig. 13.2)

A CS that addresses metal transport in plant roots: This case focuses on how gold (Au) is taken up by eucalyptus roots, transported to leaves, and stored. This case highlights several topics related to scientific experiments, hypothesis forming, and data analysis in soils and plant mineral nutrition fields.

CS 1 website link: https://www.nsta.org/ncss-case-study/fields-gold

CS 2: Farming in Space: Developing a Sustainable Food Supply on Mars (by Joyner and Allen 2015)

A CS that addresses how to grow food crop plants on Mars: This case focuses on food supply sustainability and plant nutrients for human health and well-being. This case highlights several topics related to seeds, fertilizers, soils, temperature, water usage, time to grow, and protein content.

CS 2 website link: https://www.nsta.org/ncss-case-study/farming-space

CS 3: Cancer Cure or Conservation (by Lizotte and Knapp 2003) (Fig. 13.3)

A CS that addresses the development of anticancer drug Taxol: This case focuses on medicinal plants and secondary metabolites in plants. This case highlights several topics related to ethnobotany, design of medicines, and land use in plant biology.

CS 3 website link: https://www.nsta.org/ncss-case-study/cancer-cure-or-conservation

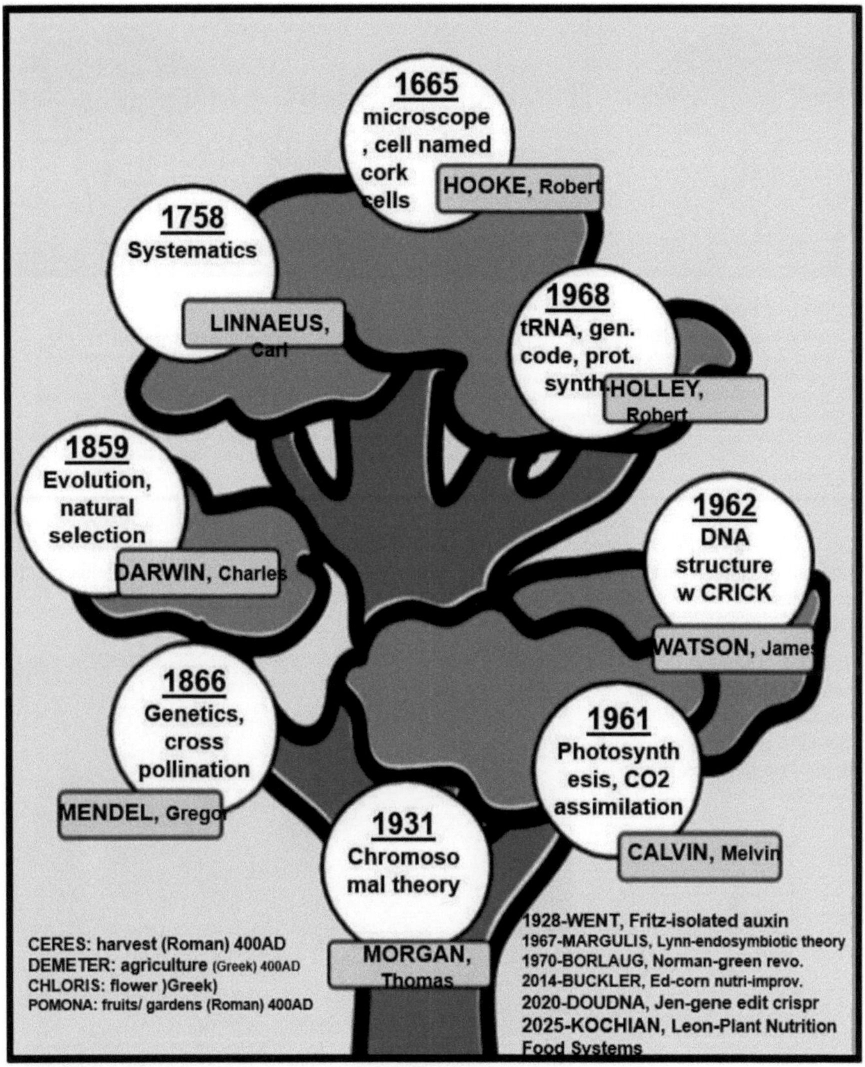

Fig. 13.1 Infographics of this chapter

CS 4: Golden Rice (by Prudhomme-Genereux 2009)

A CS that addresses vitamin A deficiency prevention by the development of golden rice. This case focuses on genetically modified organisms (GMOs) concepts around the world. This case highlights several topics related to GMO crops, nutritional quality of plants, golden rice adoption, and carotene biosynthesis pathway.

CS 4 website link: https://www.nsta.org/ncss-case-study/golden-rice

CS 5: Sweet Beets: Making Sugar out of Thin Air (by Sletten 2015)

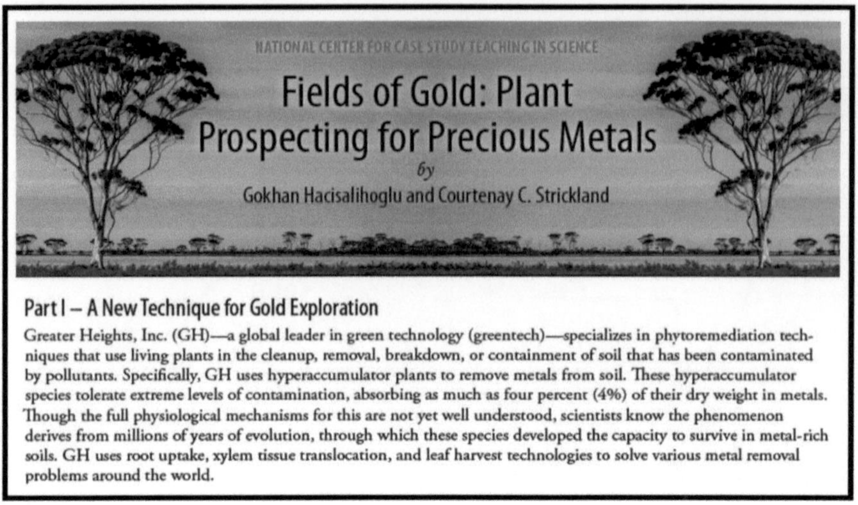
NATIONAL CENTER FOR CASE STUDY TEACHING IN SCIENCE

Fields of Gold: Plant Prospecting for Precious Metals

by

Gokhan Hacisalihoglu and Courtenay C. Strickland

Part I – A New Technique for Gold Exploration

Greater Heights, Inc. (GH)—a global leader in green technology (greentech)—specializes in phytoremediation techniques that use living plants in the cleanup, removal, breakdown, or containment of soil that has been contaminated by pollutants. Specifically, GH uses hyperaccumulator plants to remove metals from soil. These hyperaccumulator species tolerate extreme levels of contamination, absorbing as much as four percent (4%) of their dry weight in metals. Though the full physiological mechanisms for this are not yet well understood, scientists know the phenomenon derives from millions of years of evolution, through which these species developed the capacity to survive in metal-rich soils. GH uses root uptake, xylem tissue translocation, and leaf harvest technologies to solve various metal removal problems around the world.

Fig. 13.2 The 2019 case study "Fields of Gold: Plants Prospecting for Precious Metals" by Hacisalihoglu and Strickland (2019). Visit https://www.nsta.org/ncss-case-study/fields-goldfor more details

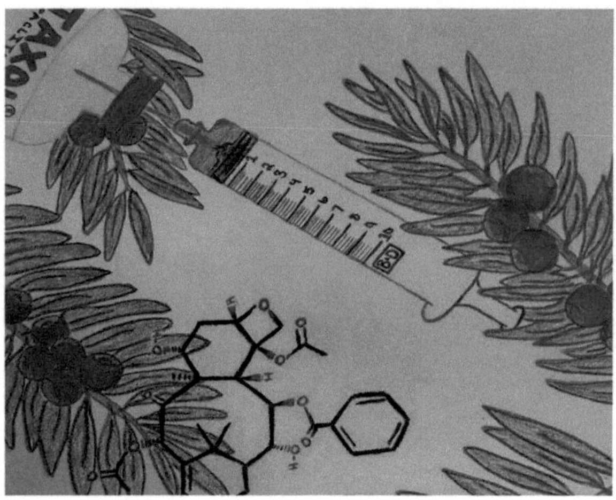

Fig. 13.3 Taxol, a chemotherapy drug to stop cancer cell division, was originally isolated from pacific yew tree bark (*Taxus brevifolia*)

A CS that introduces photosynthesis and carbon cycle: This case focuses on sugar beet plants. This case highlights several topics related to the function of roots, leaves, growing seasons, photosynthesis, and sucrose production.

CS 5 website link: https://www.nsta.org/ncss-case-study/sweet-beets-making-sugar-out-thin-air

CS 6: Corn Ethanol: Using Corn to Make Fuel (by Hartman and Eichler 2012)

A CS that addresses the biofuel production from plants: This case focuses on corn ethanol. This case highlights several topics related to biofuels from plants, renewable energy sources, and crop loss topics.

CS 6 website link: https://www.nsta.org/ncss-case-study/corn-ethanol

CS 7: A Future Without Coffee (by NSTA 2025a)

A CS that addresses plant diseases: This case focuses on coffee plants and leaf rust disease. This case highlights several topics related to plant disease resistance and protection of coffee plants.

CS 7 website link: https://www.nsta.org/playlist/future-without-coffee

CS 8: How Does Our Garden Grow (by NSTA 2025b)**?**

A CS that introduces garden plants: This case focuses on gardening techniques. This case highlights several topics related to community gardening and plant cultivation.

CS 8 website link: https://www.nsta.org/science-teacher/science-teacher-mayjune-2022/how-does-our-garden-grow

For more on this topic, see Amos et al. (1997), Diamond (2002), Hacisalihoglu (2024), Hacisalihoglu and Strickland (2019), (Hartman and Eichler 2012), Joyner and Allen (2015), Lizotte and Knapp (2003), NSTA (2025a, b), Prudhomme-Genereux (2009), Sletten (2015), USDA-NRCS (2000), YouTube Channel (2025), and many references therein.

References

Amos A, McCarthy D, Zimmet P (1997) The rising global burden of diabetes and its complications: estimates and projections to the year 2010. Diabetic Med 14:S1–S85

Diamond J (2002) Evolution, consequences and future of plant and animal domestication. Nature 418:700–707. https://doi.org/10.1038/nature01019

Hacisalihoglu G (2024) Case studies (cs) method: perspectives from learning plant nutritional biology with cs applications. Preprints, 2024071225. https://doi.org/10.20944/preprints202407.1225.v1

Hacisalihoglu G, Strickland C (2019) Fields of gold: plant prospecting for precious metals. NSTA 1–15. https://www.nsta.org/ncss-case-study/fields-gold

Hartman JD, Eichler JF (2012) Corn Ethanol: using corn to make fuel? NSTA 1–6. https://www.nsta.org/ncss-case-study/corn-ethanol

Joyner HS, Allen ML (2015) Farming in space? Developing a sustainable food supply on mars. National Center for Case Study Teaching in Science. NSTA 1–5. https://www.nsta.org/ncss-case-study/farming-space

Lizotte PA, Knapp GE (2003) Cancer cure or conservation. A Question of Health for Humans and the Ecosystem. https://www.nsta.org/ncss-case-study/cancer-cure-or-conservation

NSTA (2025a) A Future Without Coffee. Why might there be a future without coffee, and can we do anything about it? https://www.nsta.org/playlist/future-without-coffee

NSTA (2025b) How Does Our Garden Grow? Science Education as Community Gardening. https://www.nsta.org/science-teacher/science-teacher-mayjune-2022/how-does-our-garden-grow

Prudhomme-Genereux A (2009) Golden Rice. NSTA 1–9. https://www.nsta.org/ncss-case-study/golden-rice

Sletten SR (2015) Sweet beets: making sugar out of thin air. NSTA 1–5. https://www.nsta.org/ncss-case-study/sweet-beets-making-sugar-out-thin-air

USDA-NRCS (2000) The PLANTS database. National Resources Conservation Service. https://plants.usda.gov/

YouTube Channel (2025) DrhaciExplains YT Channel. https://www.youtube.com/c/DrhaciExplains

Chapter 14
10X Plant Bio Mastery: Top of the Class College Success Tips

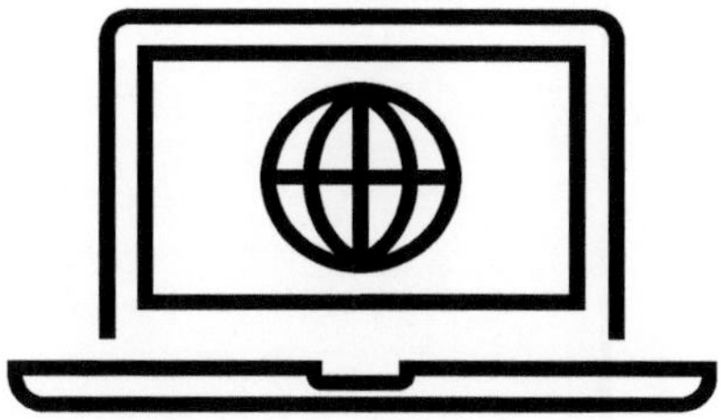

10X Plant Bio Mastery: Top of the Class College Success Tips

Contents

14.1 Big Picture: Quick Read Chapter Infographic

14.2 Student Learning Outcomes (SLOs) of the Chapter

- Explain growth mindset to develop success strategies.
- Identify study skills for mastering plant biology.

G. Hacisalihoglu, *Super Greens*, https://doi.org/10.1007/978-3-032-17047-7_14

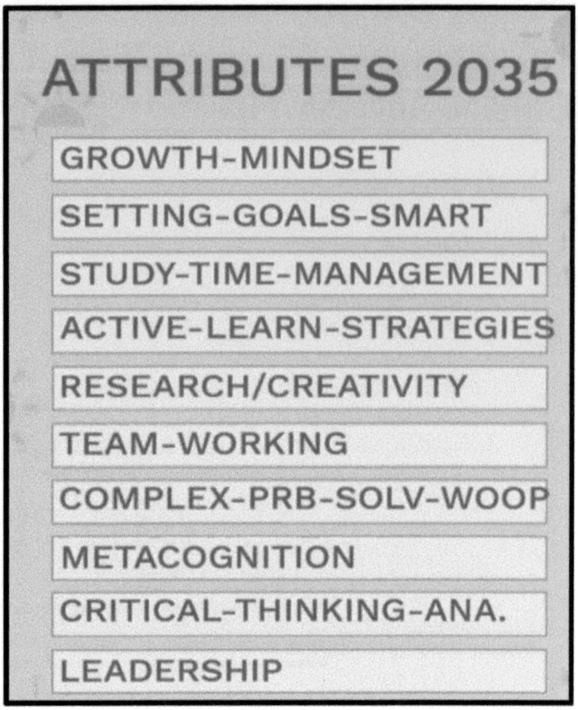

Fig. 14.1 Infographics of this chapter

- Identify new technology tools and apps for academic success.
- Identify top ten academic success skills.

14.3 Bridging the Gap

Education is a gift, and establishing a studying routine is the first step to get started. It is important to develop effective **study habits** for all undergraduate studies, including biological sciences. Moreover, frequent daily reviews of the information and course material are definitely important for **long-term memory**. During learning, our brain activates processes to absorb and apply new complex concepts. Brain's auditory cortex (in the temporal lobe) processes auditory inputs, including lectures and pattern recognition. Brain's hippocampus (in the temporal lobe) converts all those inputs to long-term memory and stores them. Consequently, the long-term memory plays a key role in the mastery of studied material.

Do you want higher grades in plant biology and graduate with a higher GPA? Then what are the 10X best study skills and tips for plant biology? **The top ten undergraduate student success tips are summarized in** Fig. 14.2**.**

For more on this topic, see Fiorella and Mayer (2015), Hacisalihoglu (2022), and many references therein.

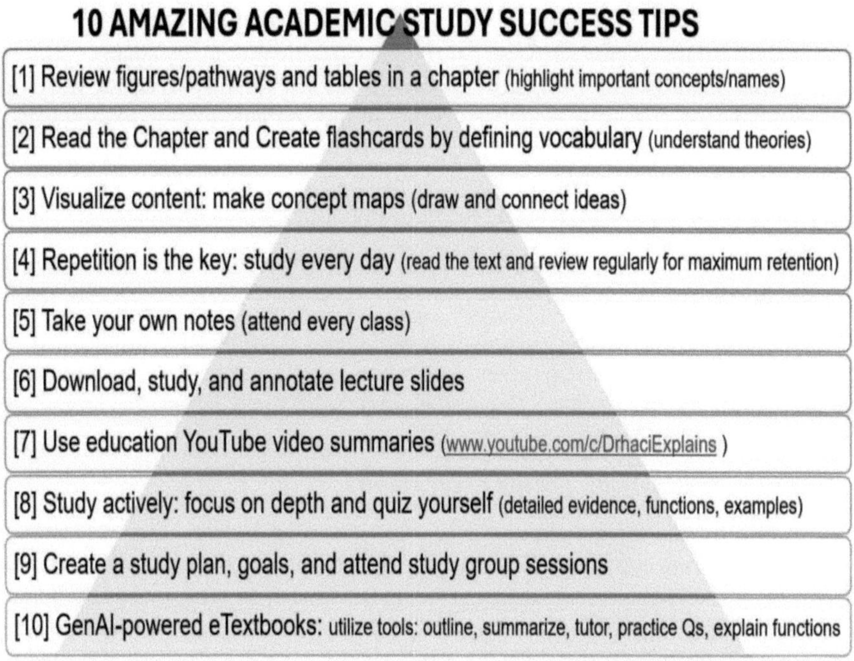

Fig. 14.2 10X academic success tips for STEM/plant biology learners

1. **Review figures/pathways and tables in a chapter (highlight important concepts/names)**

The first tip for student success is to thoroughly examine all figures and tables in each chapter of the textbook. Figures summarize complex information and provide a clearer understanding of concepts. Therefore, figures could help students identify patterns and relationships in the chapter. Always carefully review the captions, labels, units, and any additional details of each visual to highlight your comprehension. Lastly, figures could be useful study aids to help you quickly recall major points in the chapter, make connections, and enhance the retention of the plant biology content.

2. **Read the chapter and create flashcards by defining all vocabulary (understand important theories)**

The second tip for student success after reading the chapter is to create flashcards by defining key vocabulary from the chapter. Flashcards are active study tools that help retention of the content by repetition. Start by writing the vocabulary on one side of the flashcard and its definition and examples on the other side. Flashcards could also be used to quiz yourself on plant biology.

3. **Visualize content: make concept maps (draw and connect ideas)**

The third tip for student success is to visualize the big picture by creating concept maps. This method helps organize complex information by showing linkages between concepts for a deeper understanding of the chapter.

Start with the central theme at the top, and then branch out to related key concepts by using lines, arrows, and explanations. Concept maps could also help identify gaps in knowledge as well as review and recall information in plant biology.

4. **Repetition is the key: study every day (read the text and review regularly for maximum retention)**

The fourth tip for student success is repetition by reading both the text and the reviewing the chapter multiple times. Repetition helps improve long-term retention of the details in the chapter. Repetition could also help noticing patterns and connections in plant biology.

5. **Take your own notes (attend every class)**

The fifth tip for student success is to take your own lecture notes. By actively taking notes during class helps with not only active listening but also engaging more deeply with the chapter content that is being discussed. In return, active listening could help improve your ability to synthesize the material and gain a better grasp of plant biology concepts.

6. **Download, study, and annotate lecture slides**

The sixth tip for student success is to download, study, and annotate your lecture slides and the course materials. Studying and annotating the slides help students identify key concepts and figures.

Start by adding your own notes and explanations to each slide. In the end, annotated slides become study guides that you can use to efficiently review plant biology concepts.

7. **Use education YouTube video summaries** (www.youtube.com/c/DrhaciExplains)

The seventh tip for student success is to review YouTube video summaries of your lectures. YouTube video summaries are a great tool to highlight the most important concepts. Incorporating video summaries into your studying improves your ability to synthesize the material and gain a better understanding of plant biology concepts.

8. **Study actively: focus on depth and quiz yourself (detailed evidence, functions, and examples)**

The eighth tip for student success is to study actively by focusing on depth rather than just surface-level memorization. Start by exploring detailed examples, real-world applications, and connections between concepts. Active studying through depth helps students master plant biology concepts.

9. **Create a study plan, goals, and attend study sessions**

The ninth tip for student success is to create a study plan, set goals, and attend study sessions.

Start by creating a study plan and detailed calendar with specific goals. Study groups could also provide an opportunity for motivation to improve plant biology learning.

10. **GenAI-powered eTextbooks: utilize tools: outline, summarize, tutor, practice Qs, and explain functions**

The tenth tip for student success is to utilize GenAI-powered eTextbooks with interactive features like outline creation, summarization, tutoring, and generating practice questions. Taking advantage of these advanced novel tools could support improved academic outcomes in plant biology.

For more on this topic, see Baker and Hacisalihoglu (2014), Brown et al. (2014), Hacisalihoglu (2022, 2023), Hacisalihoglu et al. (2020), Hacisalihoglu et al. (2018), Quave (2014), YouTube Channel (2025), and many references therein.

References

Baker C, Hacisalihoglu G (2014) Excursions in teaching plant sciences through local ethnobotany of the food–medicine continuum. Springer, pp 245–259. https://link.springer.com/chapter/10.1007/978-1-4939-0422-8_15

Brown PC et al (2014) Make it stick: the science of successful learning, 1st edn. Harvard University Press, p 336

Hacisalihoglu G (2022) Level up college success: research, writing, and more key skills for STEM learners. ISBN13: 9798218066260. www.amazon.com/author/drhaci

Hacisalihoglu G (2023) Innovative research experiences for underrepresented undergraduates: a collaborative STEM research program as a pathway to Graduate School. BioRxiv 2023. https://doi.org/10.1101/2023.01.03.522432

Hacisalihoglu G, Stephens D, Edington M (2020) Enhancing undergraduate student success in STEM fields through growth-mindset and grit. Educ Sci 10:279. https://doi.org/10.3390/educsci10100279

Hacisalihoglu G, Stephens D, Edington M (2018) The use of active learning approach in a SCALE-UP learning space improves academic performance in UG Gen Biology. PLOS One 13(5): e0197916. https://doi.org/10.1371/journal.pone.0197916

Quave CL (2014) Innovative strategies for teaching in the plant sciences. https://link.springer.com/book/10.1007/978-1-4939-0422-8

YouTube Channel (2025) DrhaciExplains YT channel. https://www.youtube.com/c/DrhaciExplains

Chapter 15
Navigating Plant Biological Research

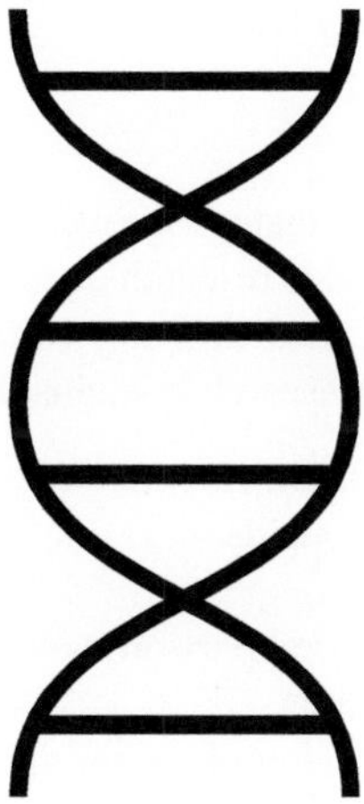

Navigating Plant Biological Research

Contents

G. Hacisalihoglu, *Super Greens*, https://doi.org/10.1007/978-3-032-17047-7_15

15.1 Big Picture: Quick Read Chapter Infographic

15.2 Student Learning Outcomes (SLOs) of the Chapter

- Explain Bloom's hierarchy of knowledge.
- Identify synthesis/creating product or proposal.
- Identify a good research question and clear hypothesis.
- Design a research experiment to test a hypothesis.

15.3 Research Idea and Solutions in Plant Biology: Real-World Impact

Plant biology researchers investigate, discover, and analyze data to solve problems. The scientific method includes research questions, forming hypothesis, experimental design, data collection, and drawing conclusions (Figs. 15.1 and 15.2). Excellent plant biology primary research contains the following research elements:

- Provides insights for an important research question and stimulates new important questions in plant biology fields
- Have a clear hypothesis
- Have a control group or other comparison metrics
- Replications or independent units
- Adequate statistical analysis (sample size, standard errors [SEM], treatment methods, and p-values
- Have complete and clear figures and tables
- Conclusions and future perspectives

DATASETS: TRAIN vs VALIDATE vs TEST-DATASET

I. **Training Dataset** (60% samples): Data to train/ teach the MODEL
II. **Validation Dataset** (20% samples): Positive control. Assess model performance
III. **Testing Dataset** (20% new samples): Final model (real-world check)→ Prediction

For more on this topic, see Dean et al. (2015), Hacisalihoglu (2022), and many references therein.

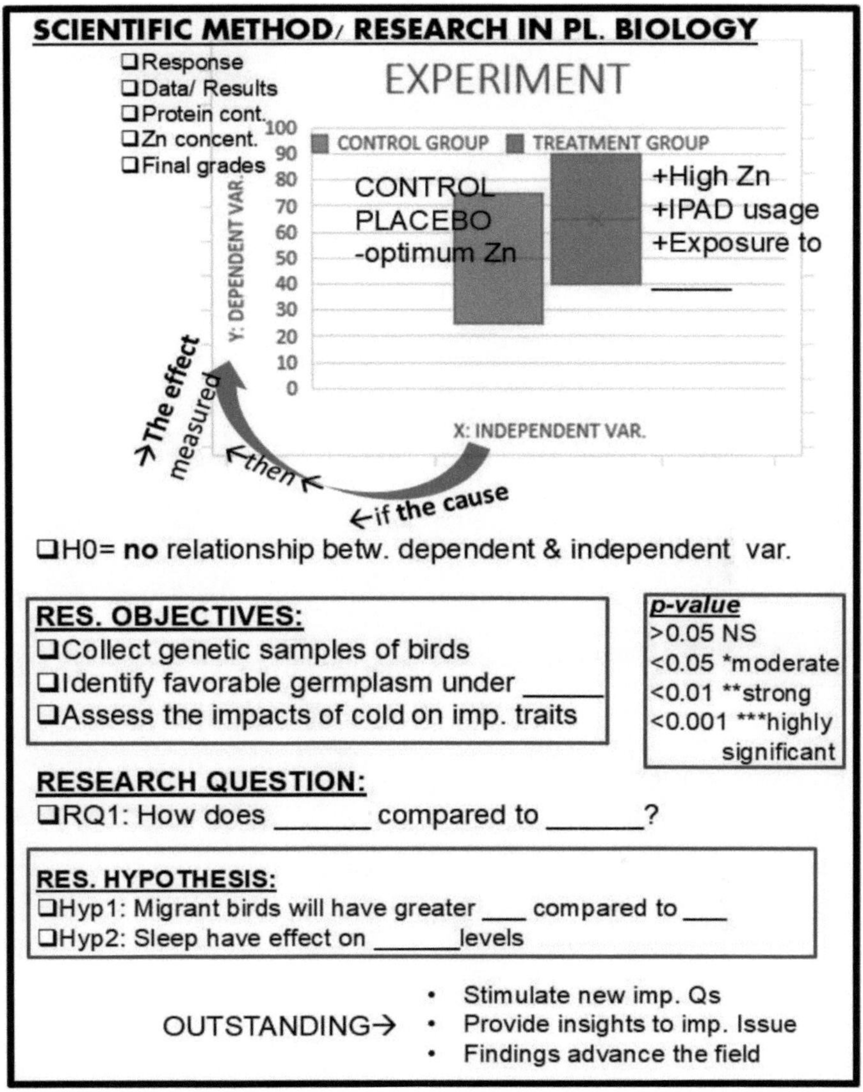

Fig. 15.1 Infographics of this chapter

15.4 Academic Presenting: Research Poster Highlights

Case Study 1 Auxin Sensitivity Is Modulated by Ammonium N in Root Growth Regulation of *Arabidopsis* Nitrogen (N, ammonium, and nitrate forms) is an essential macronutrient for plants, and, therefore, there is an increasing demand for N in global crop production. Understanding N use efficiency and its dependence on diversity of N forms remains one of the biggest challenges in agriculture. Auxins are

Fig. 15.2 Research innovation and integration of various genetics, physiological, and bioinformatics approaches in plant biology research by the author (Dr. G. Hacisalihoglu). (Image: G. Hacisalihoglu, PhD)

phytohormones involved in plant root development. The action of auxin on controlling root system architecture is known to be affected by nitrate uptake and assimilation. In this study, a quadruple knockout mutant (*qko*) of *Arabidopsis thaliana* ammonium transporters was generated in the background of DR5-GFP auxin reporter line to elucidate the cross talk between ammonium transport and auxin signaling pathway. We performed a comparative analysis of *qko* mutant and wild-type *Arabidopsis* plants that displayed different growth of primary root (PR) and lateral roots (LR) depending on N forms and auxin accumulation. We identified a

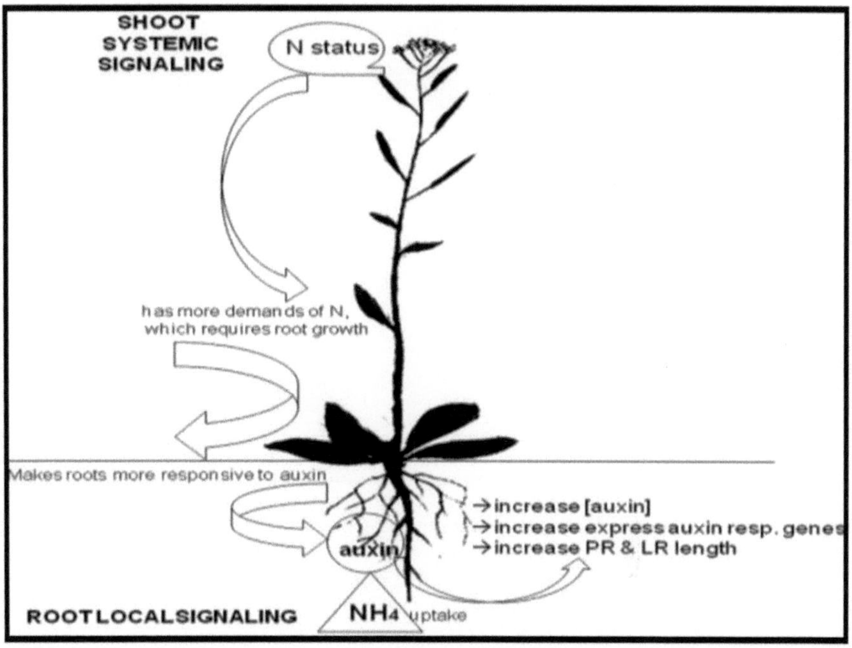

Fig. 15.3 The effect of nitrogen forms and ammonium transporters on auxin response and root growth phenotypes. Auxin sensitivity is modulated by ammonium in root growth regulation of *Arabidopsis thaliana* (Hacisalihoglu et al. 2013)

relationship between PR growth on ammonium and expression of DR5-GFP. Altogether, our results suggest a possible involvement of ammonium for modulating auxin-regulated root growth response in *Arabidopsis* through systemic signaling from shoot as well as local effect of auxin from root system. Since N efficiency is a very important agricultural trait, our findings may contribute to further improving safe plant growth under low nitrogen conditions. (Fig. 15.3, Hacisalihoglu et al. 2013).

Case Study 2 Gene Silencing [Inhibition of Anthocyanins] in Crop Plants via RdDM Targeting Recent evidence indicates that genetic and epigenetic processes play a role in gene transcriptional regulation (Fig. 15.4a). It remains unknown how epigenetic mechanisms, particularly RNA-directed DNA methylation (RdDM), cause transcriptional gene silencing in crop plants.

RdDM is an epigenetic mechanism that suppresses transcriptional gene expression. Normal anthocyanin pigments are purplish color; however, they showed reduced pigmentation to green color when silenced via RdDM epigenetics (Fig. 15.4b).

Case Study 3. Seed Nutritional Quality Traits Profiles and Metabolite Biodiversity Quality traits (texture, flavor, aroma, taste, and nutrient content) in food crop plants have received increasing attention in recent years worldwide. Crop

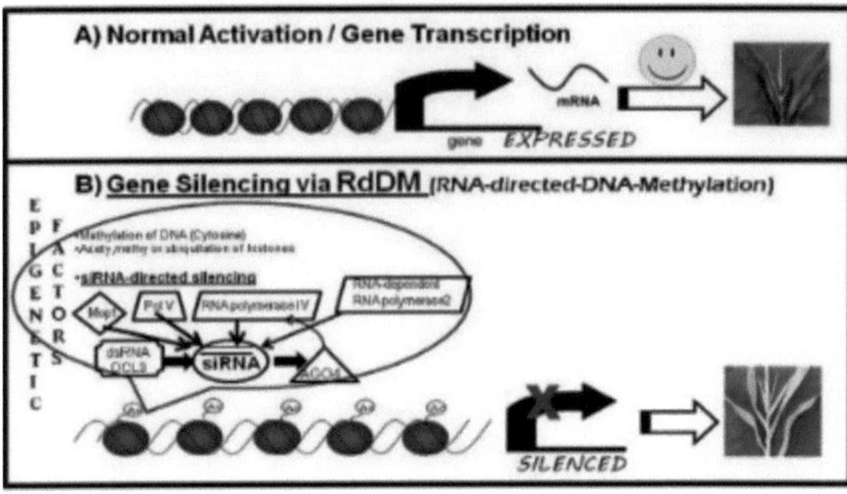

Fig. 15.4 RdDM adds methylation to DNA sequences, prevents TF binding, and therefore disrupts gene expression. (**a**) Normal transcriptional gene activation with anthocyanins (purple pigment); (**b**) transcription gene silencing via RNA-directed DNA methylation (RdDM) pathway targeting and inhibiting anthocyanins (green pigment). (Image: G. Hacisalihoglu, PhD)

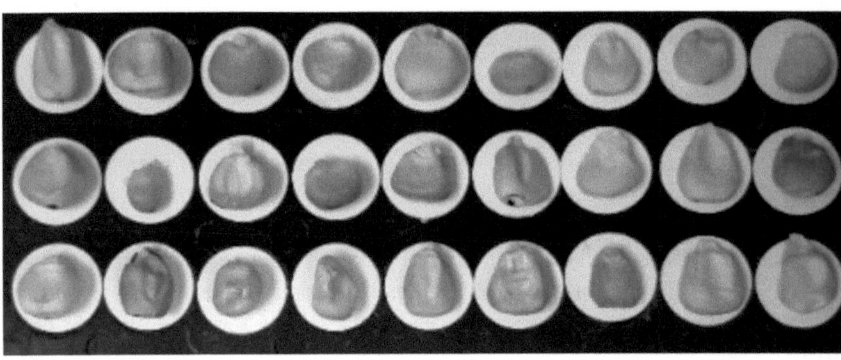

Fig. 15.5 Representative seed biodiversity of maize nested association mapping (NAM) panel with 29 genotypes, including reference parent B73, tropical, temperate, and popcorn lines. Seeds are important sources of traits of proteins, carbohydrates, lipids, macronutrients, and micronutrients for the human diet. (Image: G. Hacisalihoglu, PhD)

varieties and genotypes have great variation, and often each genotype shows a unique nutrient profile (Figs. 15.5 and 15.6). Seeds are important sources of macronutrients and micronutrients for the human diet, including zinc (Zn), iron (Fe), potassium (K), calcium (Ca), and much more. Furthermore, seeds of many food crops have the potential to deliver FDA-recommended daily value percentages (% DV) for multiple macro- and micronutrients.

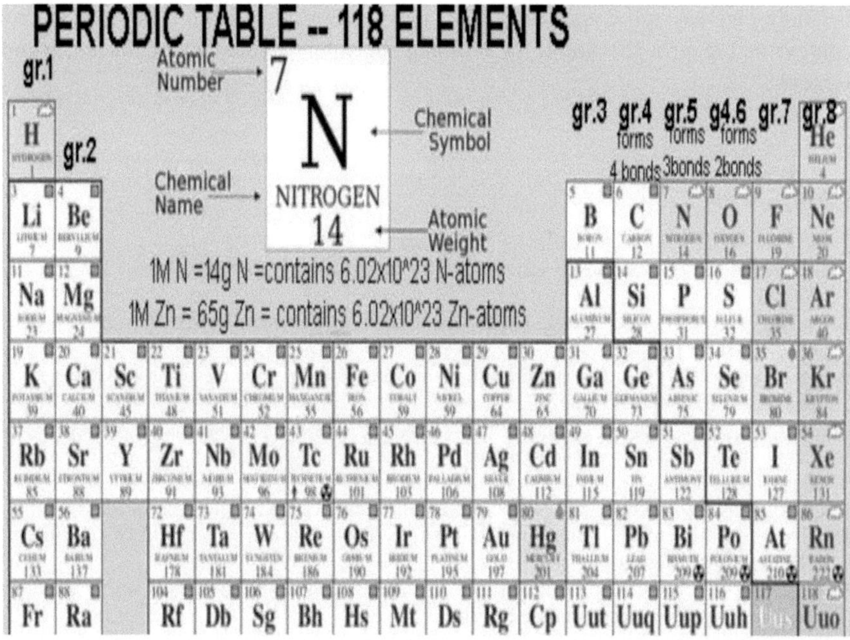

Fig. 15.6 Periodic table of elements. Essential mineral nutrients include 17 elements out of 118 total elements. P is essential for energy transfer in cells with ATP), Zn is important for immune function, and Fe is essential for oxygen transport in the blood. 1 mole has 602,214,076,000,000,000,000,000 number of atoms (=Avogadro's number 6.022×10^{23} atoms)

For more on this topic, see Altschuld (2003), Brown et al. (2014), Busby et al. (2017), Campos et al. (2021), Day (2006), FDA (2014), FAO (2017), Hacisalihoglu (2007, 2011, 2021, 2022, 2023), Hacisalihoglu and Armstrong (2022, 2023), Hacisalihoglu et al. (2013, 2018), Hacisalihoglu and Settles (2017), Hacisalihoglu and Strickland (2019), Hacisalihoglu and Ross (2010), Harshberger (1896), McGinnis (2023), Stephan (2008), Quave (2014), USDA-NRCS (2000), YouTube Channel (2025), and many references therein.

15.5 Conclusion and Looking Forward: Take Plant Bio to the Next Level

Plant biology, a core field of biological sciences that focuses on **plant form, function**, genetics, and environmental interactions, is essential for global challenges, including **future sustainable plant-based food**, medicine, low environmental footprint, and biofuels. By integrating tools from the twenty-first century, Super Greens helps faculty, students, and growers envision the future plants, as plant biology is becoming increasingly important worldwide. Gain insights into plants in the human context and stay in the know for building the future of learning with the **Super Greens!**

References

Altschuld RA (2003) US science education: The view from a practicing scientist. Rev Pol Res 20(4):635–645

Brown PC et al (2014) Make it stick: the science of successful learning, 1st edn. Harvard University Press, p 336

Busby PE et al (2017) Research priorities for harnessing plant microbiomes in sustainable agriculture. PLOS Biol 15:e2001793

Campos ACA, van Dijk WF, Ramakrishna P, Salt DE (2021) 1135 ionomes reveal the global pattern of leaf and seed mineral nutrient and trace element diversity in *Arabidopsis thaliana*. Plant J 106:536–554

Day RA (2006) How to write and publish a scientific paper. Cambridge University Press

Dean A, Morris M, Stufken J, Bingham D (2015) Handbook of design and analysis of experiments. CRC Press, Boca Raton

FAO (2017) The future of food and agriculture: trends and challenges. Food and Agriculture Organization of the United Nations, Rome

FDA (2014) U.S. Food and Drug Administration: Nutrient content claims for "good source", "high", "more", and "high potency". Code of Federal Regulations. Subpart D, Title 21, 101.54. https://www.ecfr.gov/

Hacisalihoglu G (2007) Germination characteristics of three warm-season turfgrasses subjected to matriconditioning and aging. HortTechnology 17(4):480–485. https://journals.ashs.org/horttech/view/journals/horttech/17/4/article-p480.xml

Hacisalihoglu G (2011) Plant biology concepts and laboratory manual. Kendall Hunt Publishing, Iowa. ISBN-13: 9780757589133. https://www.amazon.com/Plant-Biology-Concepts-Laboratory-Manual/dp/0757589138

Hacisalihoglu G (2021) From growing to biology: plants, 1st edn. Florida State University, p 157. ISBN13: 9781736577912

Hacisalihoglu G (2022) Level up college success: research, writing, and more key skills for STEM learners. ISBN13: 9798218066260. www.amazon.com/author/drhaci

Hacisalihoglu G (2023) Innovative research experiences for underrepresented undergraduates: a collaborative STEM research program as a pathway to graduate school. *BioRxiv* 2023(01): 522432. https://doi.org/10.1101/2023.01.03.522432

Hacisalihoglu G, Armstrong P (2022) Flax and Sorghum: multi-element contents and nut values within 210 vars & pot selection for future climates to sustain food security. Plants 11(3):451. https://doi.org/10.3390/plants11030451

Hacisalihoglu G, Armstrong P (2023) Crop seed phenomics: focus on non-destructive functional trait phenotyping methods and applications. Plants *12*(5):1177. https://doi.org/10.3390/plants12051177

Hacisalihoglu G, Ross Z (2010) The influence of priming treatment on germination and soil emergence of nonaged and aged annual ryegrass seeds. Seed Sci Technol 38(1):214–217. https://doi.org/10.15258/sst.2010.38.1.21

Hacisalihoglu G, Settles AM (2017) Quantification of seed ionome variation in 90 diverse soybean (*Glycine max*) lines. J Plant Nutr 40:2808–2817

Hacisalihoglu G, Strickland C (2019) Fields of gold: plant prospecting for precious metals. NSTA, pp 1–15. https://www.nsta.org/ncss-case-study/fields-gold

Hacisalihoglu G et al (2013) Interaction of nitrogen-forms and auxin to promote root formation as revealed by the analysis of an Arabidopsis. In: 24th International Conference on Arabidopsis Research (ICAR), Sydney, Australia

Hacisalihoglu G et al (2018) Modulation of early maize seedling performance via priming under sub-optimal temperatures. PLoS One 13:e0206861

Harshberger JW (1896) The purposes of ethnobotany. Botanical Gazette *21*:146–154

McGinnis K (2023) RNA-mediated regulation of gene expression and plant response in Maize. Annual Meeting

Quave CL (2014) Innovative strategies for teaching in the plant sciences. https://link.springer.com/book/10.1007/978-1-4939-0422-8

Stephan PE (2008) Science and the university: Challenges for future research. CESifo Econ Stud 4(2):313–324

USDA-NRCS (2000) The PLANTS Database. National Resources Conservation Service. https://plants.usda.gov/

YouTube Channel (2025) DrhaciExplains YT Channel. https://www.youtube.com/c/DrhaciExplains

Index

G. Hacisalihoglu, *Super Greens*, https://doi.org/10.1007/978-3-032-17047-7

MIX
Papier aus verantwortungsvollen Quellen
Paper from responsible sources
FSC® C105338

If you have any concerns about our products,
you can contact us on
ProductSafety@springernature.com

In case Publisher is established outside the EU,
the EU authorized representative is:
Springer Nature Customer Service Center GmbH
Europaplatz 3, 69115 Heidelberg, Germany

Printed by Libri Plureos GmbH
in Hamburg, Germany